Toward a Culture of Nature

Toward a Culture of Nature

Environmental Policy and Sustainable Development in Cuba

Pamela Stricker

LEXINGTON BOOKS
A division of
ROWMAN & LITTLEFIELD PUBLISHERS, INC.
Lanham • Boulder • New York • Toronto • Plymouth, UK

LEXINGTON BOOKS

A division of Rowman & Littlefield Publishers, Inc.
A wholly owned subsidiary of The Rowman & Littlefield Publishing Group, Inc.
4501 Forbes Boulevard, Suite 200
Lanham, MD 20706

Estover Road
Plymouth PL6 7PY
United Kingdom

British Library Cataloguing in Publication Information Available

Library of Congress Cataloging-in-Publication Data

Stricker, Pamela, 1962-
Toward a culture of nature : environmental policy and sustainable development in Cuba / Pamela Stricker.
p. cm.
Includes bibliographical references and index.
ISBN-13: 978-0-7391-2023-1 (cloth : alk. paper)
ISBN-10: 0-7391-2023-9 (cloth : alk. paper)
1. Sustainable development—Cuba. 2. Sustainable agriculture—Cuba. 3. Environmental policy—Cuba. 4. Environmental protection—Cuba. 5. Cuba—Economic conditions—1990- 6. Cuba—Politics and government—1990- I. Title.
HC152.5.Z9E54 2007
333.7097291—dc22 2007018644

Printed in the United States of America

∞™ The paper used in this publication meets the minimum requirements of American National Standard for Information Sciences—Permanence of Paper for Printed Library Materials, ANSI/NISO Z39.48-1992.

This book is dedicated to my husband, Nick, and the people of Cuba who work toward and know that *un mundo major es possible* (A better world is possible.)

Contents

List of Tables

List of Acronyms

ACAO	Cuban Organic Farmers Association
AMA	Cuban Environmental Agency
CIDA	Canadian International Development Agency
CIGEA	Cuban Center for Environmental Information, Management and Education
CITMA	Cuban Ministry of Science, Technology, and the Environment
CMEA	Council of Mutual Economic Assistance (socialist bloc)
COMARNA	Cuban National Commission for the Protection of the Environment and Conservation of Natural Resources
CREE	Centers for the Production of Entomophages and Entomopathogens
CUBASOLAR	Cuban solar energy NGO
EAN	Cuban National Environmental Strategy
ENEA	Cuban National Strategy of Environmental Education
FAO	Food and Agricultural Organization (United Nations)
FANJH	Foundation for Nature and Man, Antonio Núñez Jiménez
INIFAT	Cuban Institution of Tropical Agricultural Research
MINAG	Cuban Ministry of Agriculture
NIMBY	Not in my backyard
UBPC	Basic Units of Cooperative Production (Cuba)
UNESCO	United Nations Educational, Scientific and Cultural Organization
WCED	World Commission on Environment and Development (U.N.)
WHO	World Health Organization (U.N.)

Acknowledgments

This book has its roots in a dissertation that was completed in 1996 when Cuba was in the midst of the Special Period in Peacetime. The list of the people who helped and supported me from that time until now has grown quite long. Many inside and outside of academia have generously helped me and I am very grateful to all of those listed here as well as to those who know they also belong on that list of support.

I wish to thank California State University, San Marcos for supporting my work in and on Cuba, and in particular, the College of Arts and Sciences and the Academic Senate. I wish to extend a special thank you to *all* of my colleagues, current and past, in the Political Science Department for their support, advice, and mentorship as well as Vicki Golich, Dean of the College of Arts and Science at CSUSM. I would also like to thank Ron Chilcote of the University of California, Riverside for his tireless support, advice and encouragement of my work on this book and for introducing me to the study of development and comparative politics. He first suggested that I visit Cuba when I was a graduate student. Additionally, I wish to acknowledge the incredible support and assistance from the librarians and staff at CSUSM, particularly Teri Roudenbush.

A number of other academics have assisted me over the years and I thank them for their generosity, particularly, Keith Griffin of UC Riverside who first encouraged me to study sustainable agriculture as a graduate student; Paul Gersper of UC Berkeley who helped me make important connections within Cuba; Peter Rosset of Food First for sharing hard-to-find documents from Cuba with me early on in the research process; Daniel Faber of Northeastern University for our discussions on social justice, and Cynthia Chavez

Metoyer of CSUSM who has served as an invaluable mentor, friend, and never-ending source of encouragement, advice, and support.

I would also like to thank Don Bray, Jennifer Abbassi, Cynthia Metoyer, and Ranjeeta Ghiara for their critical readings of earlier versions of this book. Responsibility for any problems of fact or interpretation remains, of course, mine. Others who read my manuscript at different times for rather similar purposes are acknowledged with gratitude and respect for the work they have done, including Barbara Metzger and Susan MacEachern, Editorial Director at Rowman and Littlefield. I also wish to thank MacDuff Stewart and Patrick Dillon of Rowman and Littlefield/Lexington Books for their work and guidance on my book.

This book could not have been completed without those Cubans working to create a more sustainable society. I would like to extend my deepest gratitude to the Cuban scientists, farmers, ministry officials, NGO representatives, academics, and others who shared their impressions and insights with me about this transition once or in multiple interviews. I am equally grateful to my dear Cuban friends and colleagues for sharing their ideas and their time with me.

I also wish to thank my students who renew me with their energy and engagement, and I especially wish to thank my student assistants who helped me over the years, Heidi, John (Phil) Doddridge, and Blanca Ciezlinka.

Outside of Cuba, I also wish to thank my dear friends who have sustained me in a variety of ways during this journey: Lynne and Steve Shevinsky, Stephanie and Carl Hurst, Anna DePuy, Antonia and Dante Cosentino, and Claudia and Bob Nabours. I also wish to thank my family for their understanding, patience, and encouragement. Finally, I wish to thank my husband, Nick, without whose love and support, this book would not be possible.

Introduction

As long ago as 1956, geophysicist M. K. Hubbert predicted that the continental United States would reach a peak in oil production in 1970 and decline rapidly thereafter. The long-time Shell Oil employee based his forecast on models of petroleum discovery, oil extraction, and consumption. At the time, his prediction was dismissed, but later experts recognized he had been correct. Hubbert also predicted that the planet would reach a peak in oil production early in the twenty-first century. Global proven reserves of petroleum in 2000 were said to have been 1,033 billion barrels, which at the current global consumption rate of 73 million barrels a day would last for forty years (Klare 2001: 41, British Petroleum 2000: 4). The problem, though, is that this estimate works only if global oil consumption remains steady. Citing data from the U.S. Department of Energy, Klare maintains that if oil consumption rises by 2 percent a year, the existing supply will be depleted in twenty-five to thirty years (Klare 2001: 19, U.S. Department of Energy 1999: 45). Accepting even the most conservative prediction of an oil crisis in 2038, it has become abundantly clear that alternative methods for global and national energy security will have to be explored.[1]

Finding energy sources other than hydrocarbons may also be prudent in the light of their impact on the production of greenhouse gases. In February 2005 scientists from the Scripps Institute of Oceanography, the Woods Hole Oceanographic Institution, and the Hadley Center reported a "stunning" correlation between a rise in ocean temperature and atmospheric pollution, primarily caused by the burning of fossil fuels. The Global climate is seriously impacted by the amount of heat locked up in the oceans. Ocean warming can lead to glacial melting, and the subsequent addition of large amounts of colder freshwater not only causes the oceans to rise, but also affects important ocean currents. One important current that could be disrupted is the thermohaline flow, which moderates the temperature of the United Kingdom and Western Europe and the East Coast of the United States. In 2000 the Intergovernmental Panel on Climate Change announced that the release of carbon dioxide and other greenhouse gases had "contributed substantially to the observed warming over the last 50 years" and that by 2099 temperatures could be 5°C higher than in 1990, a larger increase than "the change in temperature between the last Ice Age and today" (Flavin 2001: 10). Island nations, such as Cuba, are already feeling the impact of rising ocean levels.

It is clearly time to rethink the current models of energy and resource usage based on the burning of fossil fuels because of their impact on the planet's climate. But it is also imperative to reconsider those traditional models of economic development, based on the consumption of fossil fuels, in terms of the human impact. Examination of development and resource use across the globe demonstrates over and over that poverty and resource depletion go hand in hand. According to the World Development Report 2000, economic growth has failed to reduce poverty or close the equality gap in many nations. The search for alternatives that address poverty, a declining global supply of natural resources, and environmental degradation was once part of an international debate on what was called sustainable development, but this conversation has quieted with globalization and increased economic growth around the world. It is time to revive that debate. To this end, I propose to examine how one nation—Cuba—faced a severe shortage of petroleum and managed to survive it and what lessons its survival offers for the rest of Latin America and environmentally sustainable development in human, environmental, and economic terms.

Cuba's response to its oil crisis was shaped by the collective nature of its society. In the Cuban context, rationality is not the rationality of the free market but a sense of social development within the limits of the environment. This perspective provides for development that places humans at the center and considers growth a normal and desirable progression but not at the expense of nature. For Cubans, sustainable development is found at the *intersection of social justice and environmental protection*. Development that creates disparities in the present or diminishes future generations' ability to provide for their needs is unacceptable. While clearly the socialist project has provided the conditions for development of such thinking, this mindset has evolved out of the Cuban experience itself.

Cuba has faced extraordinary challenges since the collapse of the socialist bloc, but the crisis known as the Special Period in Peacetime has been seen as something faced by all Cubans. Had individualist values prevailed, the crisis could easily have threatened the country's political and economic structure. This is not to say that individualism is completely absent in Cuba, but because of the larger social goals to which the populace is committed (for example, an end to starvation and illiteracy, lower rates of infant mortality, higher rates of life expectancy, universal access to health care, and levels of education comparable to those of the developed world) there is a greater sense of community and the common good there than in other societies. Further, while a market exists, it is neither a free market nor entirely a state-controlled one.

Cuba's economy is in a gradual transition toward a mixed economy. Wealth disparities have been greatly reduced since the 1959 revolution. Therefore, when the country faced a crisis and the government vowed to make sure that people would not starve and that as much of the cradle-to-grave social safety net as possible would remain intact, the Cuban people responded with creative ideas for contributing to the collective good. The island has one of the highest literacy rates in the world and one of the highest per capita educational levels in the Third World. It also has one of the highest numbers of scientists per capita in Latin America. While all forms of industry and occupation are viewed as working for the people, science is especially expected to work for the betterment of society. This approach is particularly evident in the cases of agriculture and energy alternatives.

In response to the events of the Special Period, Cuba left behind its proud scientifically driven Green Revolution-style of development to launch a countrywide experiment with alternative methods of agriculture. Many across the nation felt that this would be a temporary condition, an experiment that would come to an end once petroleum-based agricultural chemicals and pesticides were again easily available and affordable. On the way to this future, however, farmers and scientists began obtaining encouraging results with the new technologies, and Cubans tend to pay attention to sound scientific results. A small group of younger, more environmentally minded scientists had been researching these alternative technologies since the 1970s. But these experiments stayed in the labs. However they would soon be provided a window of opportunity to advance these new technologies. With the collapse of the socialist bloc, the country found itself facing a grave political, economic, and humanitarian crisis and at the same time, the U.S. began tightening the political and economic conditions of the Helms-Burton Act as well as its decades-long blockade. As a result, Cuba was challenged with the potential threat of grave food shortages resulting from the Soviet aid cutoff and the severe lack of petroleum products.

At the same time, as the crisis caused pressure on the island's natural resources in an attempt to satisfy needs previously met with purchased or subsidized goods, Cubans recognized the need to protect those resources. While it attempted to jumpstart its ailing economy by encouraging the development of tightly controlled enterprises involving foreign investment, the government moved to strengthen the natural resource base to preserve the country's long-term ability to sustain its population. Legislation to protect the environment and a successful campaign to raise environmental consciousness followed.

Finally, in the context of the country's isolation and its obligatory pursuit of self-sufficiency, use of sustainable agriculture and alternative energy sources intensified and gains were made in agriculture, energy conservation, and alternative medicine, drawing upon the work of those environmentally minded scientists.

My work is based upon more than a decade of research including interviews with scientists, government officials, nongovernmental organization representatives, and others as well as primary and secondary research conducted during four trips to Cuba from my initial visit in November 1994.

This book begins by describing the circumstances that set Cuba firmly on the path toward sustainable development, goes on to examine its experiences along the way, and points to the lessons of these experiences for others.

NOTE

1. While, of course, the planet will never completely run out of oil, we will reach a stage of where oil extraction becomes prohibitively expensive both in terms of the extracting and refining process as well as the technological abilities in the extraction and refining processes.

1

Sustainable Development

Poverty is not only an evil in itself, but sustainable development requires meeting the basic needs of all and extending to all the opportunity to fulfill their aspirations for a better life. A world in which poverty is endemic will always be prone to ecological and other catastrophes. —The Brundtland Commission

In 1989 following the break up of the socialist bloc, Cuba found itself in a precarious position. The country lost its major source of oil as well as an important source of loans and foreign capital, and political, economic, and social solidarity. Suddenly the island had to find a new model of development that would allow growth to occur but would also preserve the social gains it had made since the 1959 revolution. The country moved, out of necessity, first to sustainable agricultural development and from that experience, saw the effectiveness of alternative forms of development. That transition was followed by increased reliance on renewable energy resources as well as a change in mindset for Cubans regarding the environment. These changes came as a result of a sweeping framework law on environmental protection, campaigns for environmental education in the schools, universities, and public sector as well as the development of environmental ethics all the while placing humans at the *center* of such development.

Through this transition to a more sustainable society, the island has not only weathered the storm following the collapse of the socialist bloc, but has also prospered when we compare the conditions of life from the crisis known as the Special Period in Peacetime to the present. Much of that prosperity has come as a direct result of the successful implementation of sustainable models of agriculture, local food production, energy self-sufficiency, and attempts to monitor the environmental impact of development as foreign investors move to invest on the island. In this chapter, I consider how sustainable development has been conceptualized within the literature and then examine how Cuba has come to not only conceptualize sustainable development, but also implement it.

CONCEPTUALIZING SUSTAINABLE DEVELOPMENT

By analyzing one society's attempt to implement principles of sustainable development set forth in the groundbreaking United Nation's Commission on Environment and Development, the Brundtland Report, we can begin to draw out the various steps that can be taken to put these principles into practice. While the Cuban transition to sustainability continues, an examination of the case so far offers much insight into how a society can go about implementing these principles of sustainable development.

Conceptually sustainable development refers to a variety of approaches in creating a long-term environmental and economical system, which considers with varying degrees of emphasis, the needs of humans as well as the preservation of the natural resource base. Fundamentally, the problem with classifying sustainable development within categories of development theory lies in the belief that the natural world served as a limitless resource base from which humans could derive material goods. This notion underlies a number of theoretical positions ranging from neoclassical economics to even some critical approaches.

Some scholars have divided the literature on sustainable development into two distinct categories based on world views: technocentric approach, often seen in the works of neoclassical economists, to its opposite the ecocentric approach that counters the notion of viewing natural resources as resources for human use. In reality the categorization is somewhat more complicated. These approaches bracket ends of a spectrum of policy within which further delineations may be made each marking varying adherence to the dichotomous classifications. The best way to examine these concepts is to view them as a continuum with a neoclassical free market approach which embraces the use of technology to harness nature for consumption by humans at one end, and at the opposite end, a deep ecological approach is utilized which places the biosphere, not humans, at the center of their concern and thus urges the usage of the most benign technologies known to humans.

For our purposes, we will examine the portion of the literature within the larger framework that attempts to consider *both* the condition of the environment as well as the needs of people. There are certain threads that run through this portion of the literature overarched by the notion that sustainable development is a positive and desirable concept. These include first, a connection between the provision of basic needs of humans with additional needs such as freedom, equality, equity, and quality of life. A second thread that builds on the first notion calls for the active participation of those people most impacted by development. A third and related thread seeks to consider the human needs for not only the current generation, but also future generations. A fourth thread seeks to balance human needs within the limits of nature in moral, ethical, and practical terms. Finally, a fifth thread woven through the literature stitches together the social justice concerns with environmentally sensitive development.

Within the literature there is agreement that sustainable development is desirable, but there is conflict over its meaning.

Sustainable development fundamentally tackles the question of how we live our lives and how goods and resources are distributed within a society and on a global scale.

Though there is consensus that sustainable development is a desirable thing, it is much less clear as to what it actually is, how it functions, and thus how we can actually create a society that embraces sustainable development. In other words, how do we operationalize it? How do we spell out what sustainability is to discover how best to implement it in a given society?

As the pressures of global capitalism and neoliberalism intensify on a global scale, the need for sustainable development only heightens around the world. Sustainable development after all addresses environmental protection, economic and social inequality, as well as highlighting the need for popular participation in the decision making of a given society. Fundamentally, sustainable development is as Christopher Bosso argues, the only apparent legitimate (in the U.S., at least) riposte to liberal economic orthodoxy (Bosso 2005). And as Beatley and Manning argue "there is a general sense that sustainability is a good thing (and that being unsustainable is a bad thing), but will we know it when we see it?" (Beatley and Manning 1997: 3).

There is much debate on the meaning of sustainable development. The origin of the term sustainable development comes to us from the German forestry management in the eighteenth and nineteenth centuries (Lafferty and Laughelle 1999: 4, quoting Stenseth 1992a and Jacobs 1995). Pre-Brundtland usage of the term has been noted in the World Commission on Churches, World Conservation Strategy in 1980 (Donald Worster 1993), in *Building a Sustainable Society* by Lester Brown (1981), and in 1984 in *Gaia: An Atlas of Planet Management* (edited by Norman Myers). The neoclassical view holds that as production increases, human standards of living will also rise and because in their view, human needs must trump that of environmental protection, we see that their definition of sustainable development is simply neoliberal capitalist development with a thin veneer of green paint. While meeting basic needs is an important aspect, indeed central to sustainable development, it is not the only set of needs sustainable development should meet. Wyller argues, for example, that we should also promote freedom, equity, and equality along with quality of life (Wyller in Lafferty and Laughelle 1999: 11). We can conceptualize this idea in terms of the necessities of life as contrasted with the *necessities of living*. Others, such as Norwegian philosopher and deep ecologist Arne Naess, argue that we cannot call development sustainable unless it is ecologically sustainable over the "long haul" (Naess 1991: 37–38 in Lafferty and Laughelle 1999: 13). The Brundtland Report also deals with this principle, stating that development cannot impair the various systems, which support life such as the atmospheric, oceanic, and pedological (soil) systems (Brundtland 1990: 44).

William Lafferty and Oluf Laughelle correctly argue that we must not try to establish whether human needs should trump the needs of nature or vice versa, but that both human needs and the carrying capacity of nature must both be considered. They point to the importance of not valuing one dimension over another, but instead ruling out those forms of development that do not take both aspects into consideration (Lafferty and Laughelle 1999: 45). The Brundtland Report also examines this question in terms of poverty. The report calls for a reordering of needs with special priority to the essential needs of the world's poorest people (Brundtland 1990: 43). Clearly, Cuba's efforts to end starvation and illiteracy in its population as well as the government's commitment to the social development of its people despite the Special Period crisis reflects its positions on social justice within sustainable development.

Much of the difficulty within the concept of sustainable development centers on the question of whether it embraces collective equity, that is, equity between this generation and future generations (or intergenerational equity), as Amundsen argues, or the equality between members of a generation (that is, intragenerational equity), or both (Amundsen 1999: 21). When a society develops, does it consider the long-term impacts of that development? In other words, does a society sacrifice the ability of future generations to grow and develop and live a prosperous life because the development path it chose wreaks havoc on the natural resource base?

The logic here is, as we know: poverty breeds environmental degradation (this leaves to the side for the moment any moral imperative to improving the lot of those currently alive), and given that the generation currently alive depends on the environment as do future generations, we simply cannot divorce *intra* from *inter*generational equity. Indeed if we are to even attempt to have intergenerational equity, it is imperative to focus on the improvement of life for the generation now alive. Cuba has also tried to address this aspect in a couple of ways, by first providing for the basic needs of the population and then adopting and implementing far-reaching environmental protection laws, environmental education, and environmental ethics campaigns. This will hopefully shape the manner in which future Cubans interact with their environment.

The Brundtland Commission sets forth not only the fact that basic human needs are not being met around the world but that those needs and the aspirations for a better life are legitimate concerns. The commission also clarifies the connection between poverty and ecological degradation and highlights the connection between poverty and inequity in predisposing a society to ecological crises (Brundtland 1990: 43–44).

The needs for health care, potable water and sanitation, adequate nutrition, education (literacy, and social and civic responsibility), as well as favorable attitudes about the environment are asserted as necessary components of sustainable development (Brundtland 1990: 109–113). The commission does note the need for countries to address equity within the question of food security. "Inequitable distribution of production assets, unemployment, and underemployment are at the heart of the problem in many countries" (Brundtland 1990: 129). Clearly we cannot ignore the questions of international debt, export-led economies, and structural-adjustment programs often imposed by multilateral lending institutions as well as factors in the food security dilemma. The Brundtland Commission also encourages the use of land reform, and sustainable methods of agriculture, respect for the rights of subsistence farmers, pastoralists, and nomads, as well as integrated rural development to address these questions of feeding a nation's people (Brundtland 1990: 138–144). One of the first goals reached by the revolutionary government was that of ending starvation in the country, completed in part by agrarian reform and some redistribution of wealth and land, a necessary feat for the continued social development of the country in good times as well as in crisis.

In addition to the specific social justice concerns set forth already, the Brundtland Commission points to the need for a shift to dependable sources of energy. The report states that the world is in a transitional period wherein collectively we have used energy resources in an unsustainable manner and must take seriously the necessity for a sustainable energy future (Brundtland 1990: 168–169). As some scholars, such as Caltech professor David Goldstein have recently argued, the world's petroleum resources, given current global consumption patterns, will likely run out in 10–25 years (Goldstein 2004). (Goldstein refers to readily available oil cost-wise and technology-wise.) Other researchers set the end of oil date at around 2050, depending on levels of global oil consumption. Regardless of whether the end comes in 2050 or sooner, efforts must be made to respond to such a crisis across the planet. As we will see, Cuba has made much progress in its use of renewable resources in an overall transition to a more sustainable society, one that places humans *at the center* while protecting the nation's environmental patrimony for current and future generations.

Critical Interpretations

John Bellamy Foster frames the question of creating a sustainable society as "sustainable development of what?" (Foster 2002: 79). Too often, he notes, sustainable development means sustaining the status quo and he instead calls for a critical analysis of development itself. He posits as does as James O'Connor that capitalism must accumulate or die, according to Marx (Foster 2002: 80, O'Connor 1994). Further, economic growth both runs up against the carrying capacity of the planet as well as the material well-being of its human and non-human inhabitants. Arguably, as Giovanna Ricoveri points out, there is a moral dimension to the global ecological crisis. People are part of nature and "the exploitation of nature is also the exploitation of some people by other people" (Ricoveri 1993). Destroying nature also degrades relations between humans.

Foster answers that "environmental degradation is therefore about environmental justice as well" (Foster 2002: 81).

Indeed, while the focus of sustainable development has frequently been on continued growth, the human factor must be placed at the center of the analysis on sustainable development. What must be sustained is not only measured reasonable development and the natural resource base on which that development is founded, but more importantly, humanity. Therefore sustainable development must have social justice as a central theoretical component if we are to truly commit ourselves to equality for current and future generations. Social justice, as we will see, is a foundational principle not only for socialist policies of Cuba, but also for sustainable development as it is evolving.

Social Justice and Environmental Protection

Vandana Shiva and other ecological thinkers and activists from the Global South and some in the Global North are among those who believe that degradation of the environment has gone too far.[1] Shiva writes that sustainability must be based on principles of nature rather than that of the free market.[2] She argues that the free market views nature as a means for maximizing profits and accumulating capital. The focus on capital accumulation and economic growth, in Shiva's view, "renders valueless all properties and processes of nature and society that are not priced in the market and are not inputs to commodity production" (Shiva 1992). The result of this, she posits, often leads to development programs, which end up destroying or degrading natural resources as well as the means for people to survive (Shiva 1992: 216).

Shiva's solution to the dilemma comes from somewhat of a "deep ecological" position where she maintains that sustainability must be viewed within the context of nature itself. She argues that to sustain a natural system, the processes and cycles of nature must be observed. Shiva maintains that the crisis of sustainability centers on the lack of attention to these natural processes and, thus, nature's ability to regenerate itself. "In a finite, ecologically interconnected and entropy-bound world, nature's limits need to be *respected*; they cannot be *set* by the whims and conveniences of capital and market forces, no matter how clever the technologies summoned to their aid," she argues (Shiva 1992: 217, original italics).

Ronald Chilcote makes the connection between economics, social concerns, and environmental protection, arguing, "Sustainability involves economic viability, environmental sensitivity, and social responsiveness with emphasis on working with and preserving nature and ensuring that national resources are not depleted (Chilcote 2002: 340). In an earlier work, he also included concerns about future generations in the sustainable development definition (Chilcote 2000: 218).

Ted Benton adds to the debate by examining the concept of sustainable development and the normative demands of the Brundtland Report. He argues current world conditions cry out for social justice and ecological protection (Benton 1999: 224). Further he posits that given the current political and economic conditions and power relations, simply targeting growth in an effort to meet the needs of the poor while preserving nature for future generations is "simply not a feasible option." The problem lies within the tendency to subsume normative commitments on social justice and environmental protection within an "economistic reading of sustainability" (Benton 1999: 225). He urges us to return to the normative objectives of the Brundtland Commission for a reordering of society.

David Barkin adds a "Global South" and process-oriented perspective to the debate on sustainable development in his monograph, *Wealth, Poverty and Sustainable Development*. Barkin sees sustainable development as a process that must occur through the efforts of those impacted (Barkin 2000: 53). He argues production must be focused on the needs of local people through direct public participation in the process.

Taken together we can draw out some important lessons pertinent for the Cuban case, from this last group of thinkers regarding sustainable development: Foster, Shiva, Chilcote, Benton, and Barkin.

Sustainable development must:

consider the needs of the people, both basic and transformative;

pay attention to the condition and limits of nature in terms of environmental protection and management;

involve people as active participants in the development process setting forth the needs at the local level;

protect the equity between today's and tomorrow's generation;
address larger concerns about wealth and land distribution in a society as well as social justice.

Another concern not directly linked to sustainable development in the existing literature but necessary for a richer analysis is the concept of consumption.

Critical Analysis of Consumption

The Brundtland Commission points to the excessive use of resources supporting the Northern lifestyle as contributing to the degradation, arguing for the promotion of values to encourage consumption patterns within ecological bounds and "to which all can reasonably aspire" (Brundtland 1990: 44). The report notes that growth alone will not solve the problem and calls for enhancing the productive potential and "by ensuring equitable opportunities for all" (Brundtland 1990: 44). While those who seek to expand production, wealth, and standards of living have taken many of these sentiments to heart, they often have not addressed the expansion of opportunities for all through other means such as the distribution of wealth. This question of consumption will grow in importance to Cuba as it integrates into the global economy, and chapter 10 analyzes this point.

The Brundtland Report posits that as production grows and capital increases, the rising tide will enhance the living standards of all. But a fundamental question, far beyond the moral imperative of sustainable development, remains unanswered, which is whether human needs, beyond subsistence, can be met by material goods or even employment (which is a point Brundtland does address). This is an important aspect to understand for the Cuban case and to its future viability as the island becomes more integrated in the global economy. Clearly production has always been the focus of consumer-oriented economies rather than consumption. As Thomas Princen noted, "expanding the stock of available resources and spreading the wealth throughout the population carries a much lower political price tag than trying to redistribute resources" from the *have gots* to the *have nots.* He adds, "economic growth, facilitated at every turn by public policy, becomes the lubricant for civic processes of democratic planning and compromise" (Princen, Maniates, and Conca 2002: 5).

The conventional wisdom about production argues the environment as an input to economic growth rather than the actual foundation of the economy. Because the value of nature and natural resources are codified within property relations instead of being valued separately from the basis of the economy, the value of nature is cloaked. While there have been important works on the valuation of nature's services—Daly and others—this is not always a point considered in these discussions.

Nevertheless, natural resources are generally assumed to be utilized for the needs and desires of humanity. Paul Sweezy argues against the desire for goods to the detriment of human beings and the earth. "The desire for these destructive goods is manufactured . . . through the force of modern marketing (Sweezy 1989: 2).

Maniates points to one alternative to the consumption mindset that turns the concept on its head, arguing that "production is consumption—consumption of time, creativity, control over work and a sustaining personal sense of fulfillment and accomplishment" (Maniates 2002: 231).

While detractors might argue that potential negative impacts from simply reducing consumption of material goods can damage the employment outlook for a domestic (or even global) economy, other economists, such as Juliet Schor in the *Overspent American*, would disagree. Schor sees a drop in consumer spending as resulting in only small impacts on the economic job outlook for workers in the U.S. and posits that workers would buy less, work less, and enjoy life more. British economist Paul Elkin notes the subversive nature of frugality which he says the "near dictatorial" premise that humans can only be happy by consuming goods and services (Elkin 1998 in Maniates 2002).

It is necessary to examine this problem within an international context because consumption of material goods follows a global expansion of capital. Obviously market forces must create consumers and needs so as to create markets for goods produced. If we focus on creating consumers and training them to fulfill their wants and desires by purchasing goods rather than by interacting in social, creative, and intellectual activities we miss the opportunities for crafting human development based on the goals of self-actualization.

We need to move beyond simply satisfying the needs of subsistence to those of self-actualization, so we need to conceive and define *sustainable development* as development with a social justice character that combines the two—basic human needs and the ability of one to realize one's potential. In other words, to prevent the creation and multiplication of consuming societies requiring more and more resources to satisfy the "false needs" for consumer goods set forth by the industries as opposed to the tools one needs to realize one's intellectual, physical, and creative self so one can live more lightly on the earth.

Further as globalization spreads across the planet, the concomitant media and advertising that follows creates what Dipak Gyawali calls a "tyranny of expectations." As Coca Cola and McDonalds spread across the globe as well as television advertising and programming from the Industrialized North, people of the Global South are exposed to programs that glamorize the consuming lifestyle and create expectations that cannot be met. "These expectations, heightened by media and elite behavior, go unmet; an enduring sense of 'missing out' and being left behind takes root and frustration, dissatisfaction, and social stability surface" (Maniates 2003: 206). While Cubans might not have had too much exposure to this, those working in the tourism industry have been exposed to these influences. Further, as more and more goods flow into the island along with foreign currency, even average Cubans are exposed to the global seductive powers of consumption.

While this is not to say that many people of the Global South should not be lifted from poverty nor is it a paternalistic or Global North-privilege claim that others should not enjoy material possessions, it is instead a call for the inclusion of the concept of consumption, that is, materialism beyond the needs of survival and self-actualization, and a critique of false needs as opposed to the real non-material (social, creative, intellectual, physical, and perhaps even spiritual) transformative needs of humans in the cause of development.

In sum, sustainable development must be *redefined* to reflect the principles of both sustainable human development and nature. We must define sustainability within the context of nature balancing the needs of humans with those of the natural world, and to fulfill the normative concerns of humans and the protection of nature, we need to place social justice at the forefront of sustainable development. Finally, people need to participate in the decision-making and management processes of sustainable development. In the analysis of the case of Cuba as a nation that has implemented sustainable development through a wide array of sectors in society, we will see how social justice, humanistic development, and environmental protection work together to give a fresh interpretation to both the theoretical concerns and realities of crafting a sustainable society.

Throughout the process, however, the fact remains that the transition was very much shaped by the Cuban commitment to social justice and equitable distribution of wealth among its citizens. That point is quite likely a *necessary condition* for a true transition to a sustainable society, as we will see, when this case unfolds. Further the necessity for such a commitment will be underscored in full implementation of these sustainable policies. While this does not discount the usefulness of this analysis for other societies, it places on the table the necessity for social justice and socially just development for creation of sustainable societies. Furthermore, it sets forth a number of challenges to nations attempting to move toward a sustainable society, indeed even in Cuba. While admirable, the Cuban transition has not been without some difficulties. These difficulties as well as analysis of the lessons of this transition for other nations of the Global South will be taken up in the final chapter.

Cuba's Transition

The case of Cuba sheds light on how and why there was a societal-wide move toward sustainability in one country. In this way, the critical case study approaches the question of sustainable development at the *macro*level, that is, across a variety of sectors in one nation, as opposed to the *micro*level, city or regional, examples which have dominated the literature. In this way, the case study of Cuba can help fill in this theoretical lacuna. The significance of this approach is to allow for the teasing out of various aspects of sustainable development to offer insights to other nations attempting to make the same journey to a sustainable society while balancing human development with environmental protection. By examining the various sectors of society contributing to Cuba's transition, the different pieces of the puzzle come to light that are needed for such a move to environmentally sustainable development.

For Cubans, sustainable development is found at the intersection of *social justice and environmental protection*. The Cuban case is significant in both theoretic terms as well as in practical terms as it offers one roadmap to a more sustainable society. The lesson the Cuban experience demonstrates is that the concept of sustainable development can be *revitalized* in the literature and in practice by connecting these two aspects: social justice and environmental protection to sustainable development. This connection allows for the process of realizing human potential in the present and future without the destruction of a nation's environmental patrimony. This is a foundational principle of the Brundtland Report as well. The lessons also illustrate that sustainable development cannot be measured only in economically based or only in ethically based criteria. As the Brundtland Report argued, people cannot simply be viewed in terms of numbers or only as consumers. To be able to fully realize the value of humans realizing their potential, people must have their basic needs met before their intellectual, problem-solving, and creative development can commence. This point highlights the importance of the social justice commitment implicit in Brundtland and other international environmental agreements, and which informs the Cuban experience of sustainable development. Of course, the Cuban commitment to social justice preceded the Brundtland Report and its own adoption of international environmental pacts. Still this study shows that social justice serves as a necessary precondition for a society's transition to a more sustainable one. Cuba is following a path of sustainable development by a) meeting the basic needs of its citizens in food security, education, and health care; b) facilitating the development of the whole person by its commitments to education, as well as athletic and artistic expression; c) creating a wide-ranging comprehensive environmental protection legal framework; d) developing an environmental consciousness through societal-wide environmental education programs and efforts to develop an environmental ethic; e) developing self-sufficiency in food, energy, and medicine; and f) initiating efforts to enhance public participation in local-based development and planned increases in public involvement in environmental decision making and management.

The transition to sustainable development and analysis of the progress made and challenges ahead for Cuba constitute the scope of this book as well as the lessons for other nations attempting transition to a more sustainable society.

Most studies on sustainability are comparative cases of how sustainable techniques have been implemented in cities and communities (Portney 2003, Mazmanian and Kraft 1999, Cooper and Vargas 2004) yet little research has been conducted on societies trying to implement the principles of sustainable development on a *national scale*. These examples point to the importance of collaborative decision making from restoring a water basin in Wisconsin (Kraft and Johnston in Mazmanian and Kraft 1999), to changes in clean air policy implementation in Los Angeles (Mazmanian in Mazmanian and Kraft 1999), to the analysis of sustainability initiatives in a selection of American cities (Portney 2003). While analysis of these cases has not resulted in a broad theory on the implementation of sustainable development, they have made important contributions in trying to draw out commonalities between these examples of sustainable development in different settings and lay important groundwork for research in this area. My study builds upon this work by examining a nationwide transition to sustainable development across a number of sectors.

Initial examination of the Cuban transition to sustainable development led to the conclusion that this was simply a transition of necessity led by a small group of scientists, who once the socialist bloc collapsed, put forth sustainable technologies they had researched and developed, but had not seen fully implemented. While the basic concern about necessity is true, as seen in the cases of agriculture and renewable resources, by analyzing Cuba's transition to a society embracing sustainable development on a number of levels, we learn that the story is more complicated. We need to examine more factors to fully understand this transition not only to understand what happened in Cuba, but how the lessons from this experience can be applied to other nations in Latin America and the Caribbean, as well as other parts of the world.

Overview

Changes to a more sustainable society in terms of agricultural development before the economic crisis of the Special Period will be explored in chapter 2, and chapter 3 focuses on the new technologies put into place during the Special Period. In addition to changes within agriculture, Cuba also moved toward energy independence after the collapse of the socialist bloc. Chapter 4 examines where Cuba is today in terms of energy self-sufficiency and green medicine.

Chapter 5 looks at Cuba's rich and diverse natural resources while chapter 6 analyzes the legal framework concretizing the Cuban commitments on social justice and protection of the island's natural resources. In chapter 7, I examine the environmental impact of tourism and the legalistic attempts to control foreign investment. To better facilitate its transition to sustainable development, Cubans have adopted environmental education in the schools and throughout the society as discussed in chapter 8. Then, in chapter 9, I will consider what the concept of sustainable development has come to mean to a number of Cuban thinkers on the island. From this survey, I will try to distill what the Cuban perspective can offer other nations of the Global South from thinkers working to transform their societies into more sustainable ones.

In the final chapter, I analyze Cuba's transition to sustainable development from a broader standpoint. What lessons can be drawn from this example that are generalizable for other nations in Latin America and the rest of the Global South? What lessons can be drawn out that might contribute to the creation of a more environmentally conscious society in the Industrialized North? What role does a Cuban view of the "rational use of resources" and the combination of social justice and environmental protection play for the future of sustainable development? And finally, what are the challenges of population growth and consumerism for this society and others around the globe?

NOTES

1. Vandana Shiva, a physicist and an environmental activist in India, has written widely on the conflict between the free market and nature/subsistence. See Shiva's *Staying Alive: Women, Ecology and Development* (1989) as well as *The Violence of the Green Revolution: Agriculture, Ecology and Politics* (1991) for elaboration of her ideas.

2. Deep ecology, a phase coined by Norwegian philosopher Arne Naess, calls for a deeper inquiry into humans' relationship with the earth and society. This inquiry, according to Naess, will require humans to rethink their relationship with the natural world and then discover new ways in which to interact the biotic world. One of the primary precepts of deep ecology is that every life form has in principle a right to live and grow. For more on deep ecology, see Naess's classic work *Ecology, Community and Lifestyle*, 1989.

2

Politics, Economics, and the Shift to Sustainable Agriculture

Agriculture is nothing more than the application of basic biology and other natural sciences to the art of growing food. —Advisory Committee to the Rockefeller Foundation, Spring 1991.

Throughout Cuba's history, sugar has been king. Cuba has also produced tobacco, coffee, and citrus fruits. In the period before the 1959 revolution, a little more than half of Cuba's population was concentrated in the urban areas (Kay 1988: 1239). The agricultural sector was a monocrop export economy based on sugar. Sales of sugar contributed more than 25 percent of the total national income and comprised 80 percent of all exports.[1]

In the pre-revolutionary years enormous sugar and ranching *latifundias* dominated the landscape. Large landowners held more than half the land while the small farmers who operated three-quarters of the farms worked only 15 percent of the land (Kay 1988, Bianchi 1964). Export crops such as sugar, tobacco, and coffee were widely cultivated, and food for domestic consumption was imported. Most agricultural workers labored on sugar *latifundias;* however, they were only employed from December to May. The May to December period was known as the "dead season" (*tiempo muerte*). Nearly half of all agricultural workers were unemployed during this time. Normally, 10 to 15 percent of the farm workers were unemployed during the December to May period (Forster 1982, Kay 1988).

Concentration of landownership in the hands of a few guided and drove the Cuban Revolution. In his defense speech during the 1953 trial for the assault on the Moncada barracks, Fidel Castro spoke about the populace's lack of access to land.[2]

> In terms of struggle . . . we're talking about the 600,000 Cubans without work, who want to earn their daily bread honestly without having to emigrate from their homeland in search of a livelihood; the 500,000 farm laborers who live in miserable shacks, who work four months of the year and starve the rest, who don't have an inch of land to till and whose existence would move any heart not made of stone . . . 100,000 small farmers who live and die working land that is not theirs, . . . to die without ever owning it, . . . like feudal serfs have to pay for the use of their parcel of land by giving up a portion of its produce, who cannot love it, improve it, beautify it nor plant a cedar or an orange tree on it because they never know when a sheriff will come with the rural guard to evict them from it (Castro 1961).

Within five months of the revolution in 1959, the government began implementing its first agrarian reform program, a step that would eventually culminate in the United States' decision to break off its relations with Cuba (Brudenius 1981).

Under the first agrarian reform law, many landholdings over 30 *caballerías* (about 403 hectares) were appropriated by the state; tenancy and sharecropping arrangements were outlawed (Kay 1988). A caballería is a tract of land approximately 33.33 acres and a hectare is 2.47 acres. By expropriating twelve thousand *latifundias,* the state acquired control of 44 percent of the farming and ranching land (Benjamin in Brenner et al. 1991, Choncol in Delgado 1965). In contrast to agrarian reform in the Soviet Union, China, and Eastern Europe, all of which were informed by a "land to the tiller" principle, Cuba's policy focused on state farms. The estates appropriated by the state became state farms and cooperatives in line with Castro's belief that the parcelization of land would lead to a drop in overall production.

> I found that upon the victory of the Revolution that the idea of land division still had a lot of currency. But I already understood by then that if you take, for example, a sugar plantation of 2,500 acres . . . and you divide it into 200 portions of 12.5 acres each, what will inevitably happen is that right away the new owners will cut the production of sugarcane in half in each plot and they will begin to raise for their own consumption a whole series of crops for which in many cases the soil will not be adequate (Castro in Lockwood 1969).

The idea that the state farming is the superior form of production persists to a degree, although state farms are now being broken up into cooperatives based on usufruct, called "basic units of cooperative production." Whereas many of the cattle ranches were converted into state managed farms, the sugar and rice farms became cooperatives under the direction of the National Agrarian Reform Institute. The farm workers exercised some control over the cooperatives and cooperative members decided in a national vote in 1961 to convert the co-ops into state farms so that they might be able to receive higher wages, better benefits, and membership in the agricultural workers' union (Forster 1992: 2). Some private farms were also created. One hundred thousand people who had been land tenants, sharecroppers, and even squatters were granted ownership title for up to twenty-seven hectares. Peasants were also permitted to purchase an additional forty hectares (and landlords were required to sell), and therefore people in the rice and tobacco production areas of Pinar del Rio saw a significant rise in their standard of living (Forster 1982: 2).

A second agrarian reform occurred in 1963 when the state appropriated nearly all farms larger than five caballerías (sixty-seven hectares). "It is estimated that after the second agrarian reform between 80 and 85 percent of the country's land was expropriated and about 70 to 75 percent of the nation's land was in the hands of the state, while the remainder belonged to the private sector. However, the private sector had better-quality land, owning about 43 percent of the arable land" (Kay 1988). At that time, two hundred thousand private farmers on average owned plots a little less than one *cabellería*, or a little more than thirty-three acres (Kay 1988: 1241).

AGRICULTURE AND INDUSTRIALIZATION

Sugar was a hated symbol of colonialism, and in the first years following the revolution about 20 percent of it was plowed under to make way for crops for domestic consumption. At the same time, industrialization was seen as the road to development, and in 1963, Castro announced that sugar exports would provide the necessary capital (Edelstein 1981: 234). The cultivation of crops for domestic consumption had cut the foreign exchange earnings that sugar had brought in, but Cuba had saved money by reducing food imports through this diversification of agriculture. This combination of measures called the "turnpike strategy" required that agriculture be modernized and mechanized, and it was expected that the industrial and managerial skills and techniques employed in the modernization of agriculture would carry over to the budding industrial sector (Edelstein 1981: 235).

The new alliance of the mid-1960s, with the Soviet Union and other socialist countries, would eventually lead to the setting of the ten-million-ton sugar harvest target for 1970.[3] However, the state farm sector was plagued by a seasonal labor shortage due in part to the onerous task of cutting sugarcane by hand. Many former cane cutters took advantage of the different opportunities for less arduous work in the post-revolutionary years. To compensate for this lack of labor the revolutionary government attempted to fill this vacancy with mobilized workers from the urban areas.

Throughout the 1970s, mechanization of agriculture largely freed the state from having to mobilize the populace for seasonal agricultural labor.

Food production was carried out by those who worked on the state farms, cooperatives, and private landholdings. The large fields of the state farms and cooperatives were well suited to mechanized agriculture, including the use of chemical herbicides, pesticides, and fertilizers.

While Cuba did not directly participate in the Green Revolution, the agricultural technology embraced by its proponents of the Green Revolution had an impact on Cuban agriculture from the 1960s to the late 1980s. The architects of the Green Revolution utilized the tools and seeds created by modern Western science to increase food production in the so-called developing world. The new technology, involving large-scale mechanization and high-yielding seed varieties, was favored over the traditional use of animal traction and use of seeds collected from local crops and passed down generation to generation. In the historical experience of England and other countries in Europe, industrialization required the production of surplus food with fewer agricultural workers to free labor for manufacturing. It has been argued that the Green Revolution architects' mission was to transform agriculture in the Third World to reduce food shortages and undernourishment and eliminate agriculture as a bottleneck to development (Griffin 1974: 2). This would open the door to Western investment, and finally to combat the communist solution which dealt with food shortages in a much more radical manner, namely by collectivism and redistribution. To accomplish these goals, advocates of the Green Revolution drew upon the agricultural experiences familiar to them, namely, the production of corn in the American heartland (Griffin 1992). The U.S. agricultural experience was a matter of science over natural knowledge, and the founders of the International Rice Research Institute in the Philippines, where the Green Revolution was launched in Southeast Asia, believed that agriculture was simply the application of science to nature.

According to Anthony Bebbington, Rockefeller Foundation consultant Carl Sauer warned against a policy of rapid technological modernization, recommending instead that traditional peasant knowledge should be the basis for any technology diffusion program. This would have two benefits: the preservation of the culture and the utilization of the peasant farmers' agricultural experience. "The Foundation, however, saw Sauer as a type of radical indigenist whose approach would not solve problems of food supply or poverty—either rural or urban" (Bebbington 1991: 2). It proceeded with its plan, and the Green Revolution brought the American agricultural experience to the Third World. Other scholars such as Anderson, Levy, and Morrison (1991) and Boyce (1993) present similar pictures of the Green Revolution technology transfer.

Whereas conventional chemical-based agriculture is based on the short term, sustainable agricultural methods are designed to preserve the farming area indefinitely. They have been found to provide an effective alternative to the chemically based approach, which has been linked to both environmental degradation and the poisoning of humans and animals. In one of the few studies comparing conventional and sustainable agriculture in the United States, the top one-third of sustainable farmers performed competitively with conventional farmers (Journal of Soil and Water Conservation, Jan.–Feb. 1995). Sustainable or organic farmers generally spend one-fourth more labor time than do their conventional counterparts, but this extra time is spread evenly over the year (Journal of Soil and Water Conservation, Jan.–Feb. 1995: 57). Smaller farms were on the average more productive than larger ones (Journal of Soil and Water Conservation, Jan.–Feb. 1995: 57). Because of possible savings from utilizing natural inputs rather than relying on purchased chemicals for fertilization and pest control, small-scale Latin American farmers may reap more benefits from sustainable methods. Further, the collection and saving of seeds as peasants once did around the world will enable crops over time to develop natural immunities to the local pests and plant diseases.

One major constraint, often pointed to as the greatest obstacle to sustainable methods, is that the intensification needed to feed an ever-increasing population depends on increased natural inputs such as water, land, and biotic resources, with increased skill and knowledge. Low-input sustainable farming requires considerable knowledge and observation of the local environment and its particular conditions, and much of the knowledge of peasant and indigenous peoples has been lost during the implementation of "modern" agriculture. Temporal and spatial intensification is possible, though, as farmers can increase crop output by varying their planting schedules and cropping patterns to take full advantage of sun, rainfall, and nutrients across a diversity of slope and soil situations. Lowland crop production can be increased by the use of mechanized farming equipment, high-yielding hybridized seeds, and petroleum-based fertilizers and pesticides, but the chemicals required by the hybridized seeds are now being judged culpable for much of the destruction of the area's once fertile tropical soils.

"The inappropriate use of high-productivity technologies is being implicated in various forms of natural resource degradation, including nutrient loading from fertilizers, water contamination from insecticides and herbicides, and waterlogging, and salinization of land" (National Resource Council 1993: 75).

Unfortunately, the damage has not been limited to the natural environment. It is to the matter of environmental and human damage, caused by the technologies of modern chemical-based agriculture, that we now turn.

CHEMICAL-BASED AGRICULTURAL TECHNOLOGY AND ENVIRONMENTAL DEGRADATION

According to the World Commission on Environment and Development, soil erosion, desertification, deforestation, soil acidification, and contamination of water and air are the result of many shortsighted policies (World Commission on Environment and Development 1987).[4] While industrial pollution and population pressures have contributed to these problems, "agricultural policies emphasizing increased production at the expense of environmental considerations have also contributed greatly to this deterioration" (WCED 1987: 125). Approximately 65 percent of the 1.5 billion hectares of cropland across the globe has been degraded to some extent over the last fifty years (WRI 2000: 16). Soils erode during rains when their structure has been degraded from the lack of organic material that contributes to percolation and water retention.

The World Summit on Sustainable Development (WSSD) in Johannesburg, South Africa, in September 2002 pointed to land degradation as an important environmental challenge to sustainable development for this century. The commission called for action to ". . . address causes of desertification and land degradation in order to restore land, and to address poverty resulting from land degradation" (WSSD 2002: 24). Addressing land degradation would, therefore, contribute significantly to the Millennium Development Goals of reducing by half the proportion of people in poverty by 2015 and ensuring environmental sustainability (GEF 2003).

Erosion also occurs as a result of the use of heavy machinery. Soil can become compacted under the weight of the machinery, degrading the soil structure as well as the air circulation channels necessary for healthy crop root growth. Erosion is also worsened in fallow periods when the soil is left to rest without the planting of a nutrient crop. Under these conditions, a soil with a weak crumb structure becomes vulnerable to the ravages of wind and rain and simply blows or washes away. The eroded soil makes its way to other waterways, choking them with silt, thereby contributing to the flooding of other crop areas and sensitive areas such as wetlands.[5] With the destruction of the fertility of the soils, farmers then move on to other, often more fragile soils, frequently found in forest and rangeland areas, and the cycle continues. Beyond this, agricultural chemical runoff causes nitrate and phosphate contamination of drinking sources. Elevated levels of nitrates in already bacteria-contaminated drinking water sources have been associated with methemoglobinemia or blue baby syndrome, and the combination of nitrates and bacteria in drinking water has been strongly correlated with bladder cancer in the Nile Delta in Egypt and more tentatively linked to gastric cancer in Chile and Colombia (Conway and Barbier 1990: 33). Pesticides are also being examined for their effects on humans. A World Health

Organization (WHO) report estimates that one person from an "underdeveloped nation dies every minute from pesticide poisoning" (Weir and Shapiro 1981). Conway and Barbier point out that the problem is complicated by the fact that symptoms of pesticide poisoning are often mistaken for epilepsy, brain tumors, strokes, and other neurological disorders. The WHO estimates that there are more than one million acute accidental pesticide poisonings and more than twenty thousand accidental deaths per year. Additionally, there is evidence that additional pesticide-related poisonings and death are suicides (WHO 1990: 96).

Pesticide poisoning occurs mostly in developing countries. Instructions for workers using pesticides are often lacking, and where instruction is offered women and children who also work in the fields are not included. In addition, many pesticides that have been shown to be fairly toxic to humans and outlawed in parts of the industrialized world are still being sold in the developing areas. For example, I, I, I, trichloro-2, 2bis (parachlorophenyl) ethane or DDT, illegal to be used in the U.S., is made by American corporations and readily available in other countries like Mexico and Guatemala. One can also find residues of banned pesticides on produce exported from countries where those pesticides are not banned in a phenomenon known as "pesticide blowback." A study conducted in the highlands of Guatemala examined 145 nontraditional producers and found that 42 of them had been poisoned at least once (Murray 1994: 69). A University of San Carlos study of two nontraditional agricultural communities in Guatemala found that 68 percent of the farmers had experienced one acute pesticide poisoning, and another study showed half of the nontraditional producers had been poisoned at least once (Murray 1994: 69). Pesticides found as runoff in water sources have also been linked to the deaths of fish and crustaceans.[6] This problem is exacerbated by the overuse that is encouraged by government subsidies on pesticide and fertilizer sales (Repetto 1986: 64). The situation may only worsen as populations increase and land distribution ratios remain skewed (Repetto 1986: 65).

While Cuba did not participate directly with the proponents of the Green Revolution, as did many other nations in Latin America and the Global South, the technologies utilized by these movements did find their way into Cuba and the nations of the socialist bloc. The Cubans do acknowledge use of agricultural mechanization and chemicals under the schemes of the so-called Green Revolution (Cuban Ministry of Science, Technology and the Environment 2000: 1).

The proponents of the program attempted to introduce their Euro-American agricultural know-how to the South in a bid to modernize those nations to an economic takeoff point (Rostow 1960). These methods were added to agricultural landscape already altered by the changes of the colonial period, which introduced cash crops and modified patterns of land use, ownership, and water usage. The Green Revolution sought to modernize agriculture by introducing objective scientific models from the top down to replace what were seen as the primitive, backward methods used by inferior, backward people." It also sought to integrate developing nations into the global market system. The modernization of agriculture was expected to produce a more favorable balance of payments in the countries, where the agricultural sector was the major earner of foreign capital (Fletcher, Graber, Merrill, and Thorbecke 1970). Although Cuban farmers were not producing foodstuffs for export to the Western free market, they were producing goods, such as sugar, that could be exchanged for other commodities, such as oil, with their partners in the socialist bloc. Thus increasing the amount of crops produced was expected to allow the country to trade them for other goods.

Contrary to the promises made by Western nongovernmental organizations (NGOs), the Green Revolution miracle did not take place. Griffin, citing data collected from 1955 to 1970, reports that "in none of the major underdeveloped regions of the world has there been an acceleration in the growth of agricultural output" (Griffin 1974: 4). Further if data from the early 1970s is included, then one finds a "reduced trend in growth" (Griffin 1974: 4).

While some increases are noted, what provides to responsible is either an expanded cultivation area or a shift in crops from food to non-food crops as Griffin (1974) and Mellor and Desai (1973) all maintain. The Green Revolution techniques afforded the farmer considerable choice—between capital-intensive methods or labor-intensive methods; human, animal, or mechanical power; chemical fertilizers or natural fertilizers; pesticides or natural pest control; herbicides and hand weeding or no weeding; and irrigation methods or dependence on rain. These different options were not, though, equally available to all farmers. Obviously, the cost of agricultural inputs was quite high for a small farmer and low for a large landowner. Planting the high-yielding varieties without the accompanying chemical fertilizers, pesticides, herbicides, and irrigation will not produce the expected yields, at least in the short term. Developed to serve many different regions and even countries with many different ecosystems, they require a neutral environment that is created with fertilizers, herbicides, pesticides, and irrigation. At the same time, without these artificial inputs, they are quite vulnerable to the local plant diseases, lacking the naturally developed resistance of plants indigenous to the local econiches.

The new technology was supposed to provide farmers with shortened cropping cycles, which would reduce water usage overall and per plant and permit a second crop, increase employment, ready dissemination and scale-neutrality. Instead, the high-yielding varieties were tenderer than indigenous varieties and needed more care; the seeds were less resistant to flooding and drought (compared with indigenous varieties) and therefore required irrigation, and the new seeds were less resistant to insects and disease than indigenous types and so needed herbicides and pesticides. Cubans therefore had to purchase these agricultural inputs with precious capital or on credit rather than relying on alternatives derived from the land and based on traditional practices. The cutoff of these inputs sent Cuban agricultural production into a tailspin, but it also opened the door for large-scale testing of alternative, more environmentally sustainable technologies.

CRISIS AND RESPONSE

The shift from traditional chemical-based agriculture to ecological agriculture was as a result of the change in the country's political relations with the Soviet Union and socialist bloc and the economic crisis that followed, its increasingly tense relations with the United States, and the actions of a small group of Cuban scientists to help Cuba survive the crisis by utilizing their proposals for sustainable agriculture.

Although the economy experienced strong growth between 1970 and 1985, it began to slow during 1985 and 1986 and registered a small downward trend, amounting to about 4 percent, in 1990. With the breakup of the Council of Mutual Economic Assistance (CMEA), it plunged 25 percent in 1991 and 15 percent more in 1992 (Zimbalist 1993a).

Cuba obtained only $1.7 billion in imports from the USSR in 1991, a precipitous 60 percent drop from the $4.1 billion in imports the previous year. The next year only brought more difficulty as Cuba posted a trade value of 7 percent of what it had been in 1989. Cuban imports dropped from a 1989 high of $8.1 billion to less than $3 billion in 1992 (Deere 1993: 40, Deere 1994: 3).

In 1990, the country imported 8.6 million tons of oil, down 4.5 tons from the 1989 13.3-million-ton import figure, and in 1992 it received less than half of the oil, about 6.1 million tons, it had received only three years before (Deere 1993, 1994). This shortage of petroleum also hurt Cuba as the island also refined some of this fuel and then resold to other countries as an additional source of foreign capital. Enrique Pérez Marin of the Ministry of Agriculture compared the crisis of 1989 to 1990 to a similar one experienced thirty years before "In 1959 we lost our market. We lost the U.S. market. . . . We find ourselves in the same situation now, without our traditional market partner, without petroleum. We have to realize that they are both different times—different in the systems, social systems. But we had the same restraints" (Enrique Pérez Marin 1994).[7] In addition to dependency on the USSR for fuel, Cuba imported agricultural inputs, which also

Table 2.1. Cuban Imports 1989 to 1992

Item	*1989 Imports*	*1992 Imports*	*% Change*
Petroleum	13,000,000 MT	6,100,000 MT	−53%
Fertilizer	1,300,000 MT	300,000 MT	−77%
Pesticides	US $80,000,000	less than US $30,000,000	at least −62.5%
Animal feeds	1,600,000 MT	475,000 MT	−70%

Sources: Carlos Lage, *Granma*, Nov. 7, 1992; Rosset and Benjamin (1994: 22).

failed to arrive in the early 1990s. Chemical fertilizer imports dropped from 1.3 million tons to 300,000 tons in 1992 (Deere 1993).

Further exacerbating the economic situation was an increase of 18 to 20 percent in the prices of imported wheat, chicken, milk, and fuel, coupled with a decrease in its two most important export products, sugar and nickel, of 20 and 28 percent, respectively (Zimbalist 1993: 154). The consequences were devastating. "That massive shock sent the Cuban economy reeling, undermining its labor markets, spurred the rapid growth of an underground economy, and cut national income by about 45 percent between 1988 and 1992" (Zimbalist 1993). The fuel and import shortages sent shockwaves through Cuban society. Bicycles, rather than cars, soon dominated the roads. Factories reduced their work schedules or closed. The rationing program expanded to include most industrial products as well as food (Deere 1993, Mesa-Lago 1993, Bengelsdorf 1994). The agricultural sector, which was highly dependent on chemical inputs and agricultural machinery, suddenly found itself with few resources. Deere estimates that between 1986 and 1989 about 82 percent of insecticides, fungicides, and herbicides and 48 percent of fertilizers were imported (Deere 1993). Pastor places the figures for herbicides and fertilizers at 98 and 94 percent, respectively (Pastor 1992). Rosset and Benjamin maintain that the import coefficient was actually 94 percent and estimate the figures for herbicides and animal feed at 98 percent and 97 percent (Rosset and Benjamin 1994).[8]

Rosset and Benjamin correctly liken pre-1989 Cuban agriculture to that of California's Central Valley rather than that of small subsistence farms throughout Latin America. With significant deficits of artificial agricultural chemicals and the fuel to operate agricultural machinery, Cuban agriculture faced an enormous challenge. The task was to double its output to make up for the decline in imported foodstuffs with half the inputs. The country could ill afford to use scarce foreign-exchange currency to import such goods. Overall caloric and protein intake dropped precipitously, perhaps by as much as 30 percent, and there were even signs of malnutrition for the first time since the early years of the revolution. As Rosset and Benjamin put it, "Food security had shown itself to be the Achilles' heel of the revolution

(Rosset and Benjamin 1994: 4). After 1989, it was estimated that Cuba had moved from the top five Latin American nations in average caloric and protein intake to the bottom five.

TENSIONS IN RELATIONS WITH THE UNITED STATES

U.S.-Cuban relations had been deteriorating rapidly since the Mariel Boatlift that was sparked by six Cubans seeking political asylum at the Peruvian embassy in Havana in April 1980. Castro responded to the crisis by accusing the United States of manipulating the Peruvian government in its efforts to destabilize Cuba, and withdrawing the Cuban guards posted at the Peruvian embassy, a Cuban policeman having been killed during the initial embassy break-in (Castro-FCDSB May 1, 1980, Sklar 1989: 340).

Over the next two and a half days, between sevan and ten thousand people crowded into the Peruvian embassy compound, and Peru, Spain, Canada, West Germany, and the United States all agreed to take in the refugees.[9] Castro announced on April 22 that the Cuban government would comply with requests from those bringing boats from Florida to evacuate relatives in addition to refugees from the embassy. Between April and September 1980, 120,000 Cubans left.[10]

In a speech commemorating International Workers' Day in 1980, Fidel Castro said that work had already begun on plans for the country's survival of a blockade that would deprive it of food and fuel (Castro May, 1, 1980). pointing out that then-presidential candidate Ronald Reagan had expressed support for a naval blockade in Cuba. These fears only increased after Reagan's election.

Expressing concern about the strengthening of the newly organized revolutionary movement in El Salvador, the Farabundo Martí National Liberation Front (FMLN), U.S. Secretary of State Alexander Haig told the North Atlantic Treaty Organization representatives in February 1981 that the U.S. had "to deal with the immediate source of the problem [in El Salvador]—and that is Cuba."

In a white paper issued on February 23, the State Department alleged major logistical support for the FMLN from Cuba and Nicaragua (Dunkerley 1988: 385). That June the white paper assertions were discredited in the U.S. press, but in November Secretary of State Alexander Haig issued strong threats against Cuba and Nicaragua.[11]

It was in this Cold War climate that the U.S. began flexing its muscle against Cuba. In early November, the U.S. Navy began what was slated to be a month of military maneuvers in the Caribbean. Pentagon officials explained that the exercises were expected to send a message to the island

(Brenner 1989: 533). Cuba responded by mobilizing its reserves in a full alert in preparation for an anticipated U.S. invasion. In a meeting with reporters on November 16, 1981, Castro said that Cuba had to prepare for two battles, responding to an attack and fulfilling production tasks. "We must be prepared for war and for peace" (FCDSB Nov. 16, 1981). Haig and Cuban Vice President Carlos Rafael Rodríguez met secretly in Mexico, but no agreement was reached (Brenner 1989: 533). Relations remained tense between Cuba and its neighbor to the north with the tightening of the U.S. embargo through travel restrictions in April 1982 and the Cuban announcement in June that its military strength had almost doubled in response to the threatened hostilities put forth by the Reagan administration.[12]

The tension came close to a flashpoint when the United States invaded Grenada, occupying it and setting up a new provisional government in 1983. There were 784 Cubans in Grenada. Most were construction workers, and the rest were military or diplomats, depending on the source of the report (Washington or Havana). The U.S. captured 642 Cubans, killed 24, and wounded 57 (Brenner 1989: 534).[13] Cuba had previously turned down a request from Grenada, following the assassination of the Grenadian Prime Minister Maurice Bishop, to assist it militarily in the event of a U.S. attack. Castro denounced the invasion of Grenada as illegal during a memorial for the 24 Cubans killed in Grenada on November 11, 1983. He said that the United States had acted out of hatred for Grenada, and for the slain Prime Minister Maurice Bishop, in a bid to destroy the progress that the tiny island nation had made in the face of economic difficulties (Castro Nov.11, 1983). He also accused Reagan of falsely blaming Cuba for the coup and for Bishop's murder.

With the invasion of Grenada in 1983, following increasing U.S. adventurism in the region since Reagan came to power, it must have appeared that Cuba might be next, and the Soviet Union only added a new dimension to Cuban concerns by not responding to the Grenada invasion. The Cubans saw that they were on their own against the North Americans (Bengelsdorf 1994: 143, Castro Dec. 29, 1991). Castro spoke at length about this in a closing speech to the National Assembly in December 1991: "We reached the conviction that the defense of this country depended on our own skin and heart, and no one became demoralized . . . and we have to thank Reagan especially for this, . . . that in the event of a war or Yankee attack, we were the only ones who would defend ourselves" (Castro Dec. 29, 1991).

This feeling of isolation, coupled with increasing U.S. aggressiveness toward Cuba since 1981, served as the initial impetus for the rectification process that became the basis for the establishment of the Special Period in Peacetime in August 1990.

STEPS TOWARD SELF-SUFFICIENCY

Some scholars have considered the rectification process Castro's answer to Gorbachev's perestroika, but this of course is hotly contested both by Castro and other political observers. Clearly, as Bengelsdorf points out, the rectification has to be analyzed within the Cuban context. The shift in U.S. policy toward Cuba during the Reagan administration and the Soviets' lack of a response to the Grenada invasion left the Cubans with the sense of a need for self-sufficiency in many ways in defense, food, and fuel. As Bengelsdorf points out, Castro began to stress values that had been emphasized in the 1960s: "voluntarism, the moral impetus to work, the capacity of the masses to accomplish the seemingly impossible," as well as a focus on increasing efficiency in production to conserve resources (Bengelsdorf 1994: 144).

Castro realized that in the face of U.S. hostility and Soviet complacency it would be up to the Cubans themselves to see to their own survival. His approach to this situation was to revive the radical spirit of the 1960s to rally the populace for the coming struggle. In his speech to the Third Party Congress of the Communist Party in Cuba, he outlined the problems and successes over the preceding quinquennium, pointing to problems such as low productivity, budget overruns and inadequate planning, inferior product quality, waste of resources, and bloated staffs (Halbesky and Kirk 1990: xiii, Castro Feb. 20, 1986). He called for the rapid industrialization of the country, the rational use of natural resources, the reduction of imported resources, and the intensification of agriculture to permit more exports of traditional agricultural products and increasing satisfaction of the population's nutritional needs" (Castro Feb. 20, 1986).

A United Nations' Food and Agricultural Organization (FAO) report comparing Cuba's experience with those of other countries has shown poor performance in crop yield compared with the levels of chemical fertilizers and irrigation, and in a 1990 article published in *Cuba Economía,* Oscar Trinchet reported a similar finding (Trinchet 1990). Other scholars have pointed to Cuba's agricultural inefficiency in the literature, both Cuban and North American, and so did the people I interviewed.[14]

As part of the project of rectification, what was being proposed was a return to the goal of self-sufficiency pursued during the early years of the revolution. In response to the initiation of the U.S. trade embargo in August 1960, the Cuban government had reacted by developing plans for moving away from its single-crop sugar economy. But this commitment to food self-sufficiency had been abandoned when Cuba joined the Council for Mutual Economic Assistance (CMEA) and traded its sugar and later nickel ore for goods from the socialist bloc (Deere 1991, Mesa-Lago 1981). New plans for self-sufficiency were first announced during the Third Party Congress" (Deere 1993: 39, Comití Estatal de Estadisticas 1989).

The programs outlined included irrigation projects, dairy and citrus production, and a plan (called the Turquino Plan) to reforest and develop the rural mountainous areas. Production programs involving rice, poultry, pork, *viandas*, vegetables, and fruit—citrus and non-citrus were later added to the overall agricultural program (Castro 1991a: 117). *Viandas* is a Cuban word for roots and tubers, like *malanga* (manioc) and *boniato* (white sweet potato).

Emphasis on roots and tubers was seen as a way of reducing the consumption of both rice and products made from wheat flour (Castro 1991b). In 1989 Cuba imported 31 percent of its wheat, nearly 180,000 tons (Deere 1993: 38, Table 2; CEE 1989). A push for increased self-sufficiency in Havana province, where approximately 30 percent of the population lives, began in earnest with a special focus on *viandas* in September 1990 (Castro 1990, Deere 1993: 39–40). The plan was expected to cut down on fuel costs for transporting these food crops for other areas.

While the idea of the Special Period in Peacetime was first discussed in January 1990, Castro explained that the crisis period would require a drastic reduction in petroleum usage but the state would not abandon its food program.[15] In the end, Cuba faced a 3.3-million-ton dearth in oil imports from the Soviets in 1990 (Castro Oct. 1991: 27). Castro's concerns about possible fuel reductions voiced in January 1990 appeared to be coming true. The Special Period in Peacetime measures were put into place.

As Castro had pledged during his previous speeches on the possibility of a Special Period in Peacetime, food production, now referred to as the National Food Program (*Programa Alimentario*), continued with an increased fervor. The program moved from one being aimed at increasing production and decreasing food imports, as had been designed under the Rectification Process, to one that adequately fed the populace.

"The food program took on a new urgency and became explicitly focused not only on reducing Cuba's food import dependence, but also food security; assuring the population adequate levels if not the normal composition of calories and proteins" (Deere 1993: 39).

In December 1990 Minister of Agriculture Adolfo Díaz outlined the measures to be taken to reach that goal (Díaz 1990: 1–15).[16] These measures included increasing the area in which *viandas* are planted, utilizing then-idle land on state farms, cooperatives, and privately held farms, and increasing the efficiency of irrigation systems (Díaz 1990, Cardet Hernando 1991: 51). The food program also incorporated some new technologies, particularly in irrigation, such as the microjet sprinkler water system, drip irrigation, and rotational or rational grazing.[17] Díaz's report to the assembly also included alternative agricultural technologies such as crop rotation, use of cover crop green manures, interplanting, use of animal manures, and worm castings from vermiculture, many of which had been introduced on a small scale around 1986 (Cuevas 1994, Thrupp and Pérez 1989, Gersper et al. 1993).

Initially, these techniques were used on a small experimental scale and, as I will explain, were not implemented on a wide scale until after the breakdown of the CMEA, and the subsequent economic crisis, and lack of food imports and agricultural chemicals. The crisis had presented an opportunity that the advocates of these alternative technologies seized and acted upon to see them introduced nationwide.

The Cuban transition to a more sustainable form of agriculture was shaped by an exceedingly complex set of political changes and historical circumstances dating back to the early 1980s with the cowboy adventurism of the Reagan administration toward Latin America and Cuba as well as the 1983 invasion of Grenada coupled with the acquiescence of the USSR over the United States' actions. There were a number of significant political changes that occurred from the early 1980s to the breakup of the socialist bloc in 1989–1990, which have been documented in this section.

As has been argued these policy changes led Cuba to pursue self-sufficiency in food and defense capabilities, reflected in part in the Rectification Process. The food program later developed out of the food production/efficiency goals of the Rectification. Some of the approaches to this goal included alternative technologies, which we know were being researched and utilized on a small scale. The initial plans drafted by the Cubans for the "special period" ultimately enabled them to survive the most difficult political and economic crisis they have faced since the early days of the revolution, that is, the breakup of the socialist bloc from 1989 to 1990. These changes caused the Cubans to move away from the chemical—and capital-intensive Green Revolution forms of agriculture to more labor—and knowledge-intensive forms of sustainable agriculture to meet the goals of self-sufficiency.

NOTES

1. The importance of sugar on Cuban politics and economics is worth noting. The U.S. military intervened in Cuba three times in the early 1900s in relation to sugar prices and labor unrest. When sugar prices dropped during the Great Depression, the labor and student movements overthrew the dictator Gerardo Machado (who had previously been installed by the first U.S. intervention). The U.S. intervened and Dr. Ramón Grau who had been named President of the Revolutionary Government was forced to step down. The Cuban military, led by Fulgencio Batista, was put into power (Edelstein 1981). Also see Kay (1988), Acosta (1972), MacEwan (1980).

2. On July 26, 1953, Castro led 134 rebels in an attack on the Moncada barracks in Santiago in the initial attack of the insurrection against Fulgencio Batista. Castro's speech during the 1953 trial was later released as the revolutionary pamphlet, *History Will Resolve Me.*

3. The Cubans produced a record 8.5 million tons of sugar in 1970, although this fell 1.5 million tons below the 10-million-ton goal. The push had also taken a toll on the rest of Cuba's economy. Castro took the blame himself in a May 19 speech to the country.

4. The World Commission on Environment and Development, *Our Common Future*, is also knows as the Brundtland Report, after its commission chair, Dr. Gro Brundtland.

5. Another problem with the use of chemical fertilizers is the exaggeration of the problem of soil erosion. In addition to a mulching material, high levels of organic material or humus in the soil can aid in the percolation process of water by helping to build a good soil structure or a good crumb structure. A technical description is worth mentioning here at length: "A strong crumb structure minimizes the chance that soil particles will be broken loose by raindrops and maintains a high soil permeability and infiltration capacity, minimizing the likelihood that runoff will be high enough to carry away loosened soil particles. The fact that traditional fertilizers are organic is important, because application of chemical fertilizers to the exclusion of organic matter results in a weak crumb structure that leaves the soil vulnerable to erosion. A continuous supply of organic material for incorporation into the soil is therefore a critical feature of erosion control in traditional agriculture. The total yield from traditional agriculture includes, not only the consumable crop, but also the residues that serve as organic fertilizers, preventing erosion and ensuring long-term agricultural sustainability" (Marten and Vityakon 1986).

6. One study showed half of 128 fish killed off the coast of South Carolina between 1977 and 1984 were attributed to pesticides (Trim 1987, Lowe 1982).

7. Of course, during the period when Cuba was a trading partner with the U.S., the island was subject to control and dominance by its powerful neighbor to the north as analyzed by James O'Connor (1970).

8. Much of the animal feed is comprised of corn and other cereal products. Cuba's climate is amenable to the cultivation of corn; however, corn has not been widely produced. Deere can find no explanation for the Cubans failure to produce corn widely (Deere 1993)—neither could I.

9. Sklar says more than ten thousand people pushed into the Peruvian embassy while Brenner et al. place the figure at about seven thousand (Sklar in Brenner et al. 1989: 340–341, Brenner et al. 1989).

10. Sklar argues, using official statistics from U.S. government sources, that contrary to some earlier reports, the "criminal and undesirable element" among those in the Mariel Boatlift was smaller than initially reported. "Out of the total of 120,737 Cuban refugees, 1,656—a little more than 1 percent—are being held in Federal correctional institutions as 'potentially excludable' under U.S. immigration laws according to a 1980 report by the Cuban-Haitian refugee Task Force" (Sklar in Brenner 1989: 347–348).

11. For a thorough discussion of the white paper, see Dunkerley 1988, also David MacMichael, *Soujourner*, Aug. 1984. MacMichael was a CIA officer during the white paper episode.

12. Bengelsdorf says this new military strategy centered around a concept of *guerra popula*: where more than two million people were mobilized into territorial

militias. Castro often referred to the arming of the populace as proof of their support of the government (Castro in *Granma,* Apr. 8, 1990, cited in Bengelsdorf 1994).

13. In a 1991 speech to the ANPP, Castro described how in the face of increasing hostility of the U.S., the government began organizing civilian defense groups to defend the island (Castro Dec. 29, 1991, FCSDB).

14. Díaz and Muñoz (1994) argue that signs appeared in the Cuban economy that indicate stagnation in the agricultural sector between 1986–1989; they say that the yields do not correspond to the investments made in the sugarcane and non-sugarcane production. Azicri (1990) also points to deficient agricultural performance.

15. The United States' 1989 invasion of Panama to oust General Manuel Noreiga had raised the concern over increasing U.S. adventurism in Latin America (Castro Jan. 29, 1990, FCSDB).

16. Díaz in ANPP 1990: 1–15. At the time Díaz was also Vice President of the Executive Committee of the Council of Ministers.

17. Rotational or rational grazing is based on the ideas of the late French agronomist André Voisin. It is based on the notion of utilizing a series of interlocking fences to move cattle from area to area thereby avoiding overgrazing in one particular area. The pasture in a given area is also fertilized by the cattle and then allowed to go into fallow. The Cubans are also experimenting with various grasses such as guinea grass and star grass (field notes 1994).

3

From Red to Green: Scientists, Politics, and Agriculture

> *What cared the Spanish planters in Cuba, who burned down forests on the slopes of mountains and obtained from the ashes sufficient fertilizer for one generation of highly profitable coffee trees—what cared they that the heavy tropical rainfall afterwards washed away the now unprotected upper stratum of the soil, leaving behind only bare rock! In relations to nature, as to society, the present mode of production is predominately concerned only about the fit, the most tangible result; and then surprise is even expressed that the most remote effect of actions directed to this end turn out to be of quite a different, mainly, even of quite an opposite character.* —Friedrich Engels.

Sustainable agriculture had been the subject of study in Cuba among a small group of scientists for a number of years. At the first national meeting on ecology in 1980, "erosion and deforestation were identified as the major ecological problems, but pollution was expected to increase in importance. The criticism of pesticide use was raised several times but also resisted by a plant protection staff person who argued that pesticides could not be all bad since the Soviet Union produced them" (Levins 1990: 125). Cubans began to argue that the old chemical-based model had been imposed from outside and was inappropriate for Cuba. In contrast to Western agriculture, most of Cuban agriculture was conducted under collective ownership, but at the level of technology the two forms were quite similar. My interviews with Cuban scientists revealed that they considered a change in technologies to be long overdue.

The divisions between alternative and conventional agriculture seems to have been generational as well. Younger and less powerful scientists took a "green" stance while older scientists favored traditional methods (Rosset and Benjamin 1994). Scientists told me that some of the resistance to the alternative technologies came from those scientists who had been trained in the more traditional types of agriculture where heavy mechanization and use of artificial imports were seen as the only approach (Hernández, García, and Pérez Marin 1994). Further, both farmers and scientists favored chemicals, particularly pesticides, because pest problems could rapidly be eliminated (Pérez Dec. 2, 1994). Integrated pest management (IPM) techniques require careful monitoring of the fields for pest infestation as well as observation of the results of a measured application of either a biological control or a pesticide. The use of chemical pesticides is simply less labor-intensive and requires minimal special training. One problem, though, arises when the pest becomes resistant to the pesticide.

Research into alternatives to chemical pesticides increased after Castro's call for fuel efficiency and imported-resource conservation at the Third Party Congress. The program included the use of biological controls on insects, interplanting, and crop rotation to reduce pest infestations, and research on resistant crop varieties (Thrupp and Pérez 1989: 5). According to Jorge Ramón Cuevas of the Institute of Soils in Fertilizers, a small group of scientists begun conducting research in this area in the early 1970s (Cuevas, July 2000, Havana). During 1985–1986, the initial stages of the movement of alternative technologies occurred with a series of research centers on entomopathogens and entomophages (CREE).[1] The biological control methods were being used more widely, and they were being applied to nearly 465,000 hectares by 1988, though not for sugarcane (Thrupp and Pérez 1989: 7).[2] It has been estimated that the use of these alternatives to chemical pesticides saved the Cubans nearly $5 million (Thrupp and Pérez 1989: 8). At that time, however, chemical pesticides were still being imported, and they were employed as the first line of defense against pests throughout the country.

The scientists committed to ecological agriculture and other ecologists continued to promote their ideas within their work environments, political organizations, interest organizations, and when possible, through the press (Levins 1990: 126). During the late 1980s some ecologists were also promoted within the Ministry of Agriculture and elected to political office (Levins 1990, Rosset and Benjamin 1994: 75). As Richard Levins points out, "Cuban ecology activists are political, committed revolutionaries who see their struggles for ecologically sound policies as part of the duty of communists in building a new society with its own relation to nature" (Levins 1990: 126).

One Cuban scientist told me that because in the past, reports on ecological agricultural research fell on deaf ears, scientists had "placed them in the desk drawer" for later (Hernández Nov. 29, 1994). This was to change in the late 1980s and early 1990s. Roberto García Trujillo, president of the Cuban

Organic Agriculture Association until leaving the country in the late 1990s, said that Cubans had been in the process of rethinking the conventional methods: "We saw that the input system had a lot of problems. The input system did not have the results we needed" (García Nov. 22, 1994). He explained that the traditional heavily mechanized method had required the clearing of large fields that eventually led to concerns about the effects of deforestation and soil erosion. He went on to say that Cubans were becoming more concerned about the external dependence of the chemical-based model and had begun investigating alternatives: "We discovered a strong movement—not just ecologists in Cuba—all around the world, an agro-ecological movement. We studied the problem and we found there was a scientific basis for new agriculture. We thought this would be useful for our nation to move toward the new ideas and reduce our pesticides and fertilizers . . . not just for today and to feed our people, but also for the future" (García Nov. 22, 1994).

The breakup of the CMEA and the consequent shortage of petroleum and chemical agricultural inputs presented the opportunity for these scientists to push their ecological agricultural principles onto the agenda. "The Special Period created conditions to introduce biofertilizers and biological control of insects. I know people wanted to introduce it before but there was a lot of resistance," said Marta Muñoz, a social scientist at the University of Havana (Muñoz Nov. 30, 1994).[3] Enrique Pérez Marin lobbied the Ministry of Agriculture and others in academic and government circles about what later would come to be known as the "alternative model" (Rosset 1995). In 1990, the Academy of Sciences called on scientists like García and Pérez in 1990 following the Soviet petroleum shortfall, because of the research these scientists had been conducting on alternative agricultural technologies (Herrera, Hernández, and Aquilera 1994). They formed the Vegetable Biotechnology and Biological Agriculture (Bio-Ag) Front, an interdisciplinary group of scientists and technicians from thirty-six institutions to perfect such alternative technologies as biological fertilizers, plant tissue cultivation, and the biological control of pests and plant diseases and to implement the successful techniques into the field as soon as possible (Hernández Nov. 1994).

To assist in transfer of new technology from the laboratory to the field, García helped organize with the U.S.-based NGO Food First, an extension program for the alternative methods, called the Lighthouse, which included courses and conferences as well as a mobile educational center that traveled throughout the country and served as a resource for both farmers and technicians.

The technology transfer was not, however, simply imposed from the top down. Agricultural scientists were also working with the farmers in the fields, observing the techniques they used and helping them to develop solutions to farming problems (Peréz Dec. 1994). García also pointed to the need to recover agricultural techniques once used by farmers that might be useful for sustainable agriculture: "These ideas were used by Cubans years ago . . . and over a short period of forty to fifty years, we have forgotten them. We need to recover them for the sustainability of the future" (García Nov. 1994). One instance of the reintroduction of peasant technologies has been the use of the predatory ant lion (*Pheidole megacephala*) to control the sweet potato weevil (*Cylas formicarius*) (Peréz Dec. 1994, Dlott et al. 1993, Rosset and Benjamin 1994).

THE TRANSITION TO THE ALTERNATIVE MODEL

The alternative agricultural model embraced by these scientists is now being used across the island and includes biological control of insect pests and plant diseases; weed control through interplanting, weed management, and crop rotation; soil amendment and soil recovery through the application of biological fertilizers, cover crop green manures, and earthworm castings (vermiculture); and the use of animal traction.

BIOLOGICAL CONTROL OF INSECT PEST AND PLANT DISEASES

Cuba first began investigating biological control of insects, albeit on a very limited basis, in the 1930s, when the parasitic tachinid fly (*Lixophaga diatraeae*) was introduced by North American scientists to control the sugarcane borer (*Diatrea saccharalis*) (Dlott et al. 1993). More formal national research programs on biological control did not begin until much later. In 1985, after some years of research, these efforts were transformed into a major campaign, and biological control began to replace pesticides as the conceptual basis for pest management (Rosset and Benjamin 1994: 36). The CREES are local small-scale production facilities for the biological control of insects. They produce biological control products for purchase and use of local producers. Since my first visit to a CREE facility in 1994, Cuba has expanded the number of CREEs on the island and created "industrial-scale" units to serve the needs of the larger farms and cooperatives. The CREEs also investigate the use of plants in their battle to keep insect pests at bay, for example, the Neem tree.

In addition to the tachinid fly the CREEs currently use trichogramma wasps (Field notes 1994).[4] They also produce fungal and bacterial entomopathogens to control insect pests. *Beauveria bassiana* is used to control the sweet potato weevil, banana root borer, rice water weevil, and the sugarcane borer (Field notes 1994).[5] *Verticillium lecanii* kills the whitefly group of pests, which attack tomatoes, beans, sweet potatoes, and cole crops such as cabbage. *Metarhizium anisopliae* is used to kill the banana root borer; the sugarcane borer; the diamond back moth (*Plutella xylostella*), and the greater wax moth (*Galleria mellonella*). The primary bacterial entomopathogen used is *Bacillus thurgiensis*, which is used to control pests such as tomato hornworms, corn earworms, cabbage loopers, and other leaf-feeding moth and butterfly larvae (Field notes 1993).

The Cubans also utilize natural substances in their attempts to control plant diseases. For example, *Trichoderma fusarium*, a soil fungus, is used to control other soil-borne diseases such as *Sclerotium rolfsii* that affects beans and peanuts (Field notes 1994, Shishkoff 1993: 26–27). Other species of *Trichoderma* have been used on sweet peppers (Shishkoff 1993: 28). Additionally, the CREEs produce the fungus *Paecilomyces liliacinus* that attacks a certain type of nematode. Some nematodes (root-knot nematodes) injure the roots of plants such as tomatoes, potatoes, and carrots, while others attack leaves.[6] Much of the early work in the CREEs focused on biological control of the sugarcane borer. Later emphasis was placed on creating other types of biological pest control for non-sugarcane agriculture at the local level. This focus on the local conditions is a significant aspect of the alternative model. For sustainable agriculture to work, it is necessary to consider the local environmental conditions (soil types, amount of moisture accessibility, pests, weeds, and timing of crop growth).

The total cost of these biological controls is estimated to be 1.172 million pesos, approximately U.S.$58,600, whereas their chemical counterparts would set the Cubans back by U.S.$6.175 million (at time of 1994 study, the exchange rate for the Cuban peso was about twenty to one U.S. dollar) (as cited in Pérez and Vásquez 2002: 134–137). Part of the added cost of the chemicals is the necessity to use foreign capital to purchase these chemical inputs on the open market. The CREE system not only relies on the use of local currency, but also provides new employment opportunities for Cuban scientists, technicians, and agricultural engineers. For the Cubans, however, the benefits of the system go far beyond economics. They consider integrated pest management a necessary component of the urban agriculture that provides much of the fresh produce now enjoyed by city residents. If chemical pesticides had been used in these urban settings, many people could have been exposed to these hazardous chemicals. Further, the Cubans report that the use of these strategies has reduced the need for pesticides further by allowing native natural pest enemies to repopulate.

INTERPLANTING AND WEED MANAGEMENT

Weeds are frequently a more difficult problem in the fields than insects (Field notes 1994). The interplanting of cover crops to shade out weeds, crop rotation, and tilling with multiplows, which attacks weeds without turning the soil, are all techniques for dealing with this problem. (The multiplow works by lifting and dropping the soil separating weeds from their root systems.)[7] The Cubans have also established a weed database to track past experience with weeds for better control.[8] Crop rotation is also being used to prevent plant diseases and to balance the nutrients used by different crops.

SOIL AMENDMENT AND RECOVERY

The research being carried out on soils is focused on the short-term improvement of production through soil conservation and reclamation, the use of biological fertilizers, and nitrogen fixation. At the soils center I visited, a bacterium had been discovered that metabolized insoluble phosphorus and made it available to plants.[9] Another aspect of the current soil technology is the effort to rehabilitate soils. Much of the damage has come by erosion and depletion of organic matter, and increased salinity. Salinization of the soil is a result of Cuba's being a narrow island vulnerable to the ocean's salt spray and irrigation, and efforts such as the growing of halophytes (salt-tolerant) plants in this area have not had much success.[10] Efforts to reduce the acidity of the soils through remineralization efforts, especially the application of limestone and additional organic matter into the soil, have been more successful. Measures to increase the biomass of the soil include the use of biological fertilizers (*Azobacter* and *Rhizobiums)*, green cover crop manures (such as *Sesbania* and the velvet bean), and animal manure and worm castings.

Azotobacter is a nitrogen-fixing microorganism that is applied to either the soil or to the leaves of plant to provide nitrogen to plants that do not produce their own. In addition to providing nitrogen to the plants, the microorganism has been found to reduce the production time by nearly three weeks in the field, and to cut down on blossom-drop (Rosset and Benjamin 1994: 57). The Cuban researchers estimate that they have achieved a 30–40 percent increase in yield for corn, cassava, rice, and other vegetables as a result of its use of *Azotobacter* (Gersper et al. 1993: 19). In 1986 they used 0.7 tons of the inoculants, and production of the microorganism jumped to 23 tons and by 1994, they were producing 30 to 35 tons (Hernández Dec. 1994). One official estimated that the country has saved U.S.$118 per hectare by not purchasing artificial fertilizers and producing more (Hernández Dec. 1994).

Worm castings are produced by mixing partially composted animal waste (usually cow, although pig and sheep manure are also used) with leftovers from sugarcane production, coffee pulp, plantains, and municipal garbage.

With a bedding ingredient (usually a heavily carbon substance like wood shavings or paper), soil is mixed with twenty thousand California red hybrid earthworms per square meter (Cuevas 1994).[11] The redworms double in population every two to three months depending on climate conditions. Redworms are used as the earthworms found in the soil are generally of a different species that cannot tolerate the heat produced during the composting process. However, in tropical Cuba, these redworms require some respite from the heat so the piles are placed under mango and banana trees (Field notes 1994). The vermicompost beds are watered frequently by nearby sprinklers to maintain an even moisture level and to regulate their temperature. The bacteria the earthworms digest break down fresh organic material, which is added every few days, and in turn they deposit their nutrient-rich castings at the bottom. When the bed reaches a height of three feet, in about three months, the earthworms are separated from the castings by thinly spreading the material on a large flat surface for sorting, and the castings are either used in nearby agricultural fields or dried and bagged for later use. A portion of the earthworms are used to "seed" a new bed, while the remainder will be used as chicken or fish feed (Cuevas 1994).

Worm castings contain large amounts of organic matter and a variety of minerals and plant nutrients that, when added to the soil and crops as a top dressing, increase overall plant health, help control plant diseases, and improve the soil's ability to take in water and air. According to the Ministry of Agriculture, four tons of vermicompost can replace forty tons of cow manure per hectare of tobacco and results in as much as 36 percent improvement in yield, according to MINAG figures (Rosset and Benjamin 1994: 60). Further, "A 40-kg. bag of Cuban worm humus can sell for as much as $US80 to 100 on the international market, though humus production had not reached levels that permit significant exports" (Rosset and Benjamin 1994:60, MINAG-Agro-Video 1992). In an interview in Havana province in 1994, Jorge Ramón Cuevas told me, "My vision is 350,000 tons of worm castings. . . . That is our potential. Theoretically, we could produce five million tons of castings without using city waste" (Cuevas 1994).

In their efforts to find alternatives to compensate for the lack of fossil fuels, the Cubans have discovered that occasionally an old technique is more effective than a new one. Animal traction—the use of oxen to pull plows—is a traditional method that is now experiencing a revival in Cuban fields. Tillage operations had dropped from a national average of ten to twelve times per growing season to three to five times by 1992 (Gersper et al. 1993). While tillage is useful for aerating heavy clay soils and reducing weeds, excessive tillage can damage the soil composition over time, making it vulnerable to erosion. Use of oxen has an additional benefit: unlike tractors, they do not get bogged down in mucky wet soil (Field notes 1994). Further, as one animal traction specialist told me, it takes only one-third of a hectare of land to feed an ox. The reintroduction of what was a "peasant" technology is significant. Valuation of scientific knowledge and the agricultural experiential knowledge of former peasants will help strengthen the emerging sustainable agricultural model by expanding the overall knowledge base and allowing farmers and agricultural researchers to choose what works best.

SUSTAINABLE AGRICULTURE AT THE GRASSROOTS

As Cuba adopted sustainable agricultural technologies on a large scale, many considered them temporary measures, and if the U.S. blockade had suddenly ended this might have proved correct. Scientists were convinced, however, if the technologies were put into practice, it would become obvious to policy makers and ordinary Cubans that they were both economically and environmentally sustainable. As part of the food program and decentralization process currently underway in Cuba, an alternative model has grown out of the traditional backyard or kitchen-garden vegetable growing and small animal rearing that existed in cities even in the pre-revolutionary period (Companioni and Ojeda 2002: 221). The urban farmers of the *Organipónicos* have learned to build soils that contain about 50 percent humus with on-site traditional compost piles, animal manures, and vermicompost. Chickens are often raised in part of the garden and fed leftover vegetative matter. Now these raised beds, built with concrete forms and irrigated with the Israeli-produced microjet drip irrigation systems, regularly produce two to three crops per season. Crops grown in these urban settings range from green beans and okra to more traditional Cuban fare such as cassava, papaya, and white sweet potato. One of the most important crops grown, according to Gilberto Navarro Alvarez, director of the *Organopónico de Alto Rendimiento*, is a spinach-like plant, called *Espinaca baracoa* (*Talinum triangulare*). This plant contains high levels of protein, the amino acid lutein, and is considered very important in Cuba's efforts to overcome retinitis pigmentosa, an eye disease that increased with the dietary privations of the Special Period.

Companion planting is also practiced in the *organipónicos*. The pungent odor of basil is believed to prevent insect pests from finding certain plants, and therefore varieties of it are planted along the ends of the raised beds. Marigolds, said to repel some root nematodes, as well as having a strong pest-repelling odor, are also planted with crops. An added bonus from these plants comes when they flower and attract bees. Trap crops such as corn that will be used for animal feed are sometimes planted to lure insects from another more desirable food crop such as green beans. *Organipónicos* produce not only food crops but also herbs for cooking and medicinal purposes. Chamomile, Cuban oregano, basil, and *yerba buena* (a type of Cuban mint) are among the herbs interplanted with flowers in one garden. These herbs are sold to the public, as are the food crops. Off to the side, near the chicken coop, there may be a beehive that serves a dual purpose, as a home for the necessary pollinators for the garden and as a factory for bee pollen and other bee products sold to area hospitals.

Havana's popular gardens, which had always existed in small numbers, became more common once collapse of the socialist bloc was underway. An urban agriculture movement began when the Agricultural Group of the City of Havana was organized in mid-1994 with the support of the Ministry of Armed Forces, the Ministry of Agriculture, the Cuban Communist Party, and the provisional government (Caguax Communications MINAG 1998).[12] By 2000, the urban agriculture sector had created more than 160,000 jobs (López 2000: 2). The *organipónicos* are overseen by a municipal urban farm enterprise that administers and provides services to the different urban agriculture entities in a given territory (Companioni and Ojeda et al. in Funes et al. 2002: 222). These units offer extension and technical help, facilitate urban farmer and home gardener networking, as well as link participants in urban agricultural with research, educational, and service centers (Ojeda as cited in Companioni and Ojeda 2002: 222). Urban agriculture gardens provide fresh perishable foodstuffs to urban consumers, draw upon the concentrated population of the city for labor, and offer a visual respite for urban dwellers from trash dumps and vacant lots (Companioni and Ojeda 2002: 222–223).

During the Special Period years of the early 1990s, there all too often was a shortage of fresh vegetables in the capital due to a lack of petroleum products for transporting food from the countryside to the city. When I first visited Cuba in 1994, agricultural markets were offering only the traditional *viandas,* like *yuca* and *malanga*, and garlic and there was little meat available for purchase. In July 2000, my visits to the markets showed something very different. Thanks in part to the development of food self-sufficiency programs in Havana, there was a variety of fresh vegetables, fruits, meats, and even houseplants for sale at markets throughout the capital. In my visits in January and May–June 2003, the offerings at the farmers' markets had increased even more than I had noted in July 2000, even accounting for seasonal change. There were also more agricultural markets around the city as well.

Clearly producing vegetables and fruits locally enables the populace to have access to fresh food without the need for refrigerated transport, saving energy, time, and spoilage.

The urban agriculture movement consists of five groups working together to provide food to the local people through a rational use of resources: Agricultura Urbana, which oversees urban agricultural management, Horticola Metropolitana, which promotes vegetable growing, Empreso Cultivos Varios, which promotes the staple crops enterprise, Asocación Ganadera, which deals with livestock rearing, and Mi Programa Verde, which organizes community members, students, and workers to plant trees (for fruit and lumber) throughout Havana (Agro-Vide MINAG 1998).

Workers in urban agriculture grow vegetables, fresh and dried herbs, ornamental plants and flowers, fruits, dryland rice, plantains, white sweet potatoes and cassava, beans, corn, and sorghum (Rodriguez et al. 2003: 5–11). They raise bees, poultry, rabbits, pigs, sheep and goats, bovines, and aquaculture fish. There are programs to provide compost for use as fertilizers, seeds for use in cultivation, and information on irrigation and drainage, and the environment.

Those who are employed in *organipónicos* receive 50 percent of the profits of the enterprise, which can amount to four to five hundred additional pesos (which was a very high wage when my research was conducted). Part of the reason for the profits generated is that, in contrast to the former state farms, *organipónicos* do not have to sell all their products to the state.

A study of urban agriculture in Havana, by the Cuban environmental NGO Foundation for Nature and Humanity, concludes that it has played a large role in facilitating food security in the capital, and that its commercialization has helped to strengthen the local economy.[13] It points out that many institutions and enterprises that once fell under government control are now being directed to be self-sufficient confirming what my interviewees recently told me.

The Cuban Ministry of Agriculture reported that in 1999 vegetable production in intensive gardens and *organipónicos* yielded about 215 grams per person per day of fresh vegetables and herbs (MINAG 1999 as cited in Companioni and Ojeda 2002: 228). Now nearly 90 percent of Havana's produce is grown in these urban agricultural settings (Buncombe 2006). Popular participation in the urban agriculture movement in Havana has grown substantially from 3,966 in early 1995 to 26,604 in 1999 (González, Ojeda, and Pozo 1999: 18). According to a 2003 report by four representatives of the Ministry's Institution of Fundamental Study of Tropical Agriculture (INIFAT), it has generated more than 326,000 new jobs nationwide while providing much of the daily caloric intake of Cubans (Rodriguez et al. 2003: 2).

The INIFAT study concludes that food production can occur in urban settings without harming the environment, thus preserving for future generations the ability to feed themselves in a "healthy and productive environment" (Rodriguez et al. 2003: 11).

Ordinary Cubans can get help with sustainable agriculture through a kind of agricultural extension called *tiendas consultados agropecuarios.* There are fourteen of these offices in the city of Havana, and they dispense free advice on growing crops in backyard gardens, soil conditions, and biological control of insects and plant diseases. They are staffed by agronomists and other trained agricultural workers who will visit residents' gardens to diagnose plant problems and suggest solutions. Because the program is self-sustaining, they sell information pamphlets, gardening tools, and even biofertilizers such as worm castings and mycorrhizas, seeds, and biopesticides such as Bt and *Beauvaria bassiana*. What is of particular interest here is that all the methodology employed is organic or part of the sustainable agriculture model.

Because of the prevailing economic and political conditions following the collapse of the socialist bloc, Cuba put into place sustainable agricultural technologies on a large scale. These technologies were believed by many to be temporary measures that would be discarded when the crisis ended.

As Cuban scientists told me in 1994, if the crisis conditions lasted long enough for the population and government officials to witness the results of what environmentally benign technologies can actually do, then the psychological and economic hurdles against sustainable methods could be overcome. Indeed, that is exactly what occurred. The Cubans have not only implemented sustainable agriculture methods in their fields, but have also expanded the model to include a expansive urban agricultural movement which employs large numbers of Cubans and provides much of the daily vegetables and fruits for the table. They have created and expanded the use of local solutions to local problems, such as seen in the example of the CREEs, and created an industry of biological pest control. Through these efforts, they not only avoid the use of petroleum-based pesticides which have shown to be harmful to humans and the environment, but also require large amounts of precious foreign capital to purchase on the global market. Other Cubans viewed the changes occurring in the early 1990s as "temporary measures" which would be turned aside once the crisis of the Special Period ended. However, once these sustainable methods demonstrated how successful they could be, they were not only adopted across the country, but also became part of Cuba's environmental framework law. The impact of the accomplishments of the sustainable agricultural transition though does not end here. These successes also helped shape the path for other experiments and later successes as Cuba turned toward other modes of self-sufficiency. I will turn to these successes and experiments in the following chapters.

NOTES

1. Centros Reproductoras Entopatologos. Entomopathogens are bacterial, fungal, or viral diseases that attack insects. Entomophages, often referred to as predator insects, are insects that attack or parasitize other insects that are seen as pests. For example, the brachonid wasp injects its larvae into the tomato hornworm. As they mature, the larvae of the wasps will feed on the hornworm, killing it.

2. By 1988, there were 165 CREE centers.

3. Muñoz currently heads an environmental agriculture education program at a former state farm in Havana province. The program is coordinated through the Latin

American Faculty of Social Scientists (FLACSO) Cuba at the University of Havana, where Muñoz is on the faculty.

4. The tachinid fly, a large dark gray fly, also attacks cutworms and corn borers among other pests. They parasitize the pest by laying eggs in the pest that feed on and eventually kill the "host" pest.

5. The trichogramma wasps also parasitize other insects by laying eggs into pests such as the tomato hornworm, borers, cutworms, and especially important for Cuba-, the tobacco budowrm (*Heliothis virescens*) and the cassava hornworm *(Erinnyis ello*). Field notes, CREE, Havana province, 1994.

6. Interestingly, there are two other groups of nematodes, which are helpful to agriculture. They are the beneficial nematodes that help break down organic matter in the soil, and parasitic nematodes that attack pests such as squash vine borers, carrot weevils, and cabbage root maggots. Cuba is also experimenting with three varieties of parasitic nematodes: *Neoplactana* sp., *Heterohabditis* sp., and *Steinerneema* sp. For more on the production of nematodes, see Dlott et al. 1993: 12.

7. For more, see Dlott et al. 1993: 13.

8. For more on the weed monitoring system, see Rosset and Benjamin 1994; Dlott et al. 1993.

9. Oxisol and Ultisol soil types are found in certain areas of Cuba. These soils have a high content of aluminum and iron oxide, which when combined with phosphorous, inhibits the absorption of Po4 by plants; the bacteria frees the phosphorous for the plant's use. Field notes, Cuba, Nov. 1994.

10. For more, see Gersper et al. 1993: 18.

11. The earthworm now used is a hybrid of the two types of redworms (*Eisenia foetida* and *Lumbricus rubellus*) originally used in the first experiments in 1986. For more, also see Rosset and Benjamin 1994: 59–62. The worms are applied in heavy concentration to stave off attack from planaria, a natural enemy.

12. Others have argued that the movement arose spontaneously, see Murphy 1999: iv.

13. The study was carried out by the Antonio Núñez Jiménez Foundation of Humanity and Man (FANJH) with cooperation of the Cuban Ministry of Agriculture. The project leader was Maria Caridad Cruz.

4

Pursuing Energy and Medical Alternatives

Power is political.—Bruno Hénriquez, one of the founders of the solar energy power NGO, CUBASOLAR

When the revolutionary government gained power in 1959, only 56 percent of the Cuban population had energy from the nation's energy network (Bérriz 2000: 1). By the late 1980s, however, the island was supplied with enough energy to meet the needs of 95 percent of the population, and fuel consumption had increased from 0.5 ton to 1.5 tons of fuel per person annually (Barris and Madruga 2000: 4). Then when the socialist bloc collapsed, industry was nearly paralyzed, with a 35 percent drop in the gross domestic product and transportation came to a screeching halt (Barris and Madruga 2000: 5). The energy crisis sparked an increase in the use of wood and even materials from old buildings as energy sources. Cuba's access to cheap foreign petroleum had depended on the exchange of a ton of Cuban sugar for a ton of petroleum products. This arrangement obliged Cuba to produce large amounts of sugar and the agricultural inputs were petroleum-based. Producing more sugar to get more energy-producing oil required the mobilization of substantial labor for the sugar harvest.

Alternative energy sources began to be explored as a result of the worldwide energy crisis in the 1970s and the U.S. economic blockade. Significant expansion of this activity seems to have begun early in the Special Period. In November 1992, Cuban leaders initiated what came to be called a national energy development plan (Bérriz 2000).[1] The politics of power and energy took on a new meaning in the Cuban content. To any nation, particularly one in the Global South, the question of energy is very much a political issue. Self-sufficiency in energy is nearly as important as self-sufficiency in foodstuffs, particularly when the supply of sources of fossil fuels is rapidly diminishing on a global scale. While a mixed blessing given the large amounts of environmental degradation caused by the combustion of fossil fuels, knowledge of the declining reserves has sparked somewhat of a resource race among nations to secure individual energy futures as much of modern economic development and modernization is prefaced upon access to cheap fossil fuels. No nation wants to be dependent on others for the fuel that feeds the fires of economic development. Petroleum products have been Cuba's energy sources of choice for electrification of the country and transportation of goods and people. During the Special Period, Cuban per capita energy consumption, formerly about eight barrels of oil, was halved. (By comparison, the United States uses the equivalent of fifty-nine barrels of oil per person annually [Lippman et al. 1997]). The reduced oil supply led to widespread power shortages and blackouts as energy generation from oil-fired facilities dropped nearly 30 percent in 1990 (Lippman et al. 1997). Similarly, to the processes that occurred within agriculture in changing over to alternative methods of production and pest and disease control, Cuba began the process of utilization of alternative forms of energy, including solar, biomass, biogas, and wind power. Ninety-five percent of the island had been connected to the country's power grid by 1989. In response to the crisis, efforts are underway to generate power for the remaining 5 percent with renewable resources. Electrifying rural areas with renewable sources preserved the environment and proved to be more cost-effective than expanding the traditional power grid. Similarly to the agricultural case, Cuba's pursuit of environmentally sound methods brought environmental and economic rewards.

THE DEVELOPMENT OF RENEWABLE ENERGY SOURCES

Similarly to the transition in agriculture, the Cubans turned to nongovernmental organizations to facilitate the transition to renewable energy. The Cuban Society for Promotion of Renewable Energy and Environmental Respect (CUBASOLAR) was founded in 1994 to study and promote the use of alternative renewable sources of energy and for the respect of nature (CUBASOLAR 2000). It conducts research on renewable energy resources, educational efforts with children and adults, career training and development, and cooperative projects with other scientific organizations and sponsors, and it participates in academic and technical conferences to disseminate alternative energy research findings (CUBASOLAR 2000). CUBASOLAR has studied alternative energy sources, such as biomass, biogas, solar, hydroelectric, and wind.

BIOMASS

Since the Cuban economy has relied on sugarcane as its main industry for centuries, the country has reduced energy usage is to reclaim part of the vegetative leftovers from sugarcane processing. Some sixty thousand tons of sugarcane must be processed to obtain seven million tons of sugar. Extracting sugarcane juice leaves 17,500 pounds of cane waste or bagasse. It is also sometimes added to livestock feed, especially for pigs (Barris and Madruga 2000: 8, Henriquez June 2003). The burning of the bagasse is estimated to have produced roughly 30 percent of the country's energy. More than one hundred factories were involved in this process (Barris and Madruga 2000: 8). In the middle 1990s, there was a shift in the use of the sugarcane waste, bagasse, as an energy source. Energy efficiency research at the Universidad de Oriente resulted in a new process to create energy in a much more environmentally benign and energy-efficient fashion (Bérriz 2000).

Other biomass products burned for energy include wood sawdust and shavings, coffee pulp, coconut husks, and rice hulls, and the resulting ash is fertilizer (Barris and Madruga 2000: 9, Henriquez 2003). By using these waste products, the Cubans gain energy and reduce the amount of waste that would require disposal. In addition, they are utilizing fuels created by the decomposition process that occurs naturally in food and animals wastes and in wastes from the production of food and alcohol. Again, using agricultural and industrial wastes provides energy and reduces the need for waste disposal.

BIOGAS

Animal manures diluted with water and left to ferment produces methane gas that can be harvested and used for fuel. The leftover manure, more beneficial for use as a natural soil supplement than raw manure, is then spread on agricultural land. Thousands of cook stoves using this form of biogas have been installed in schools, health centers, and industrial centers around the country (Barris and Madruga 2000: 10). At the local household level, manure is collected from backyard chickens and pigs and placed in a small biogas fermenter. The methane produced is pumped into the kitchen for household use. A residential biogas fermenter costs the equivalent of U.S.$20. Clearly this is an example not only of inexpensive but also decentralized energy self-sufficiency. Biogas is also created in large-scale systems. One large-scale biogas project is the Soroa Pig Farm and Biogas Project. At full capacity, fueled by the waste from 15,000 pigs, the project produces approximately 182.5 tons of gas annually (Stowell 2001). Biogas centers opened in Villa Clara in 1993.

SOLAR

Cuba's strategic location in the Caribbean Sea has historically served as a source of difficulties, including the disruption by Columbus, exploitation by the Spanish, and as America's pawn in the Cold War. However, Cubans have recently begun to see their island's location in a somewhat different light. Most of the country is situated at 23° north latitude to the Tropic of Cancer, and it receives solar radiation of approximately five-kilowatt hours per square meter per day year-round (equivalent to about half a kilogram of petroleum) (Barris and Madruga 2000: 8). The fact that the country's width does not vary more than 3° in latitude from the widest to narrowest portion is an advantage in this regard (Barris and Madruga 2000: 7). Solar energy was used in some areas of the country before the Special Period, in solar collectors for heating water; this effort was halted when it became impossible to obtain the materials needed from abroad. In 1993 two companies, EcoSOL and Rensol, began producing flat solar collectors. The collectors were imported, and their manufacture was completed in Cuba. By attempting to coordinate these efforts, CUBASOLAR is enhancing the effort to create completed manufactured solar energy collection equipment.

Industries are also investing in solar energy systems as they recognize the savings that can be generated by switching from conventional sources. One industry group that has expressed interest in these systems is tourism. The tobacco industry has also discovered the usefulness of solar energy in the drying and curing of tobacco. Agriculture is benefiting not only from the use of solar energy collectors but also from the use of special light to maximize crop production. Special greenhouses are outfitted with tubes containing a variety of liquid compounds that filter out particular forms of light shielding the greenhouse from the heat but allowing in specific rays from the spectrum that contribute to the growth of particular crops (Henriquez 2003). Solar dryers are used to preserve food and also to dry wood for use as lumber.

Solar panels have been installed in medical clinics, social centers, and schools around the country. There has been a specific focus on the use of solar energy in Guantánamo Province, formerly the region least connected to the energy grid. In 2000 the government funded a program to electrify the primary schools throughout the island that were not already connected to the traditional power grid. More than 1,900 schools were electrified with photovoltaic panels, inverters, controllers, and batteries (Stone 2001–2002). Some of the schools also utilize a hybrid wind power/solar electric system. Each solar setup generates eight hours of electricity for a school. By June 2001 more than two thousand schools in mountainous and rural areas had been electrified for the first time with solar technology. The next phase of the solar electrification project will provide access to computers for all schoolchildren in the country, and there are plans to add one hundred thousand homes to the grid over the next five years (Henriquez 2003).

The shift to solar energy is also helping to create new opportunities for business and industry. For example, the steel industry is building a hybrid (solar/traditional-power) system (at Santa Clara) and solar heaters (at Morón) (Barris and Madruga 2000: 21). The development of the industry, similar to the CREE and *organipónico* systems in agriculture, are not only beneficial in terms of job creation and business expansion, but also it reinforces the use of renewable resources and creates products that can be exported to other countries in the Global South.

A large hurdle so-called developing nations has faced is affordable access to new forms of technology to move toward use of renewable sources of energy. With the growth of these domestic industries, the Cuban case shows how one country can create the means for renewable energy systems, as well as serves as a trading partners for others in their development of renewable energy systems.

HYDROELECTRIC

Just as Cubans are taking advantage of the island's specific geographic location in producing solar energy, efforts are underway to harness some of the waterpower produced by the country's rivers. Cuban river levels vary considerably with the rainy and dry seasons and of the main rivers only the Toa has sufficient power to be harnessed for energy. Dam construction for hydroelectric power often comes at a high social and environmental cost in displacement of populations and environmental destruction, and indeed, an attempt to construct a huge dam on the Rio Toa had to be abandoned in 1997. One of Cuba's environmentalists and revolutionary heroes, the late scientist Antonio Núñez Jimenez led efforts to halt the dam (Association of Cuban Geographers 1998). (His organization also organized meetings to raise awareness of water and marine pollution, in the Caribbean in 1994, and Havana's Almandares River, in 1996 and 1998.)

Hydroelectric resources have not been readily available given the nature of Cuba's rivers. Some hydroelectric resources are used in the Sierra Maestra, but because of the season changeability of flooding/drought, most of Cuba's rivers are not able to be utilized in this manner, at least on large-scale projects.

From an environmental and social justice perspective, small-scale hydroelectric projects from existing sites might serve Cubans better as these projects coincide with the physical nature of many of Cuba's waterways, and they do not require large investments of infrastructure, nor does their construction necessitate the relocation of the surrounding population. The local-based self-sufficient solution appears to be the most suitable.

WIND

Harnessing energy from the wind is another option in its bid for self-sufficiency. Windmills have traditionally been used to pump groundwater for cattle and other livestock. Research centers around the country are currently examining how this technology can be updated and improved upon. Wind energy is seen as a particularly useful mode for short-term use and use in remote locations. One study identified a number of sites suitable for collecting wind energy, including Cayo Sabinal, Punta de Maisí, Cayo Guillermo, Cayo Santa María, the island of Turiguanó, and Cabo Crúz in Granma Province. Additionally hybrid sources such as wind/diesel are being explored; a system installed at Cayo Romano has the potential of producing fifty-kilowatt hours per day (Barris and Madruga 2000: 12). In June 1999 the first wind park on the island of Turiguanó opened with two wind generators of 225-kilowatt hours per day each (Bérris ibid). Again we see the lessons of small-scale locally designed self-sufficient modes of renewable resource energy production working well in the example of hydroelectric power generation in Cuba.

ENERGY CONSERVATION

Efforts are underway to utilize technologies to reduce energy use in hotels, restaurants, offices, and some private homes. Compact fluorescent light bulbs are being used and energy efficiency is also being incorporated in the design of new buildings and renovation of old ones. Bioclimatic architecture entails designs that take advantage of climate and environmental conditions to best maximize comfort and minimize energy use within the building. This trend in global architecture goes beyond the passive solar design that became popular worldwide in the 1970s.

In Cuba, people working within the government as well as relevant nongovernmental organizations are working on more environmentally friendly design for construction. These groups incorporate design principles regarding window placement and selection for ventilation, natural lighting, and heating. Consideration of factors such as lighting, heating, cooling, and energy-efficiency are not only taken into account when designing a new building, but also in the renovation of older buildings which is going on to a large degree in old Havana. So we see a broad embrace of these principles across the spectrum of construction on the island.

Pointing to the adaptation of living organisms to their environment, architecture professor Dania González Couret argues that the best architectural solution will always be specific to the surroundings in which the building is placed and the way those conditions will change over time (Gonzalez 2000).[2]

In the industrial sector, there are experiments with co-generation, both in the use of waste heat to heat warm water or of gases released in the burning of fuels to generate energy.

Energy conservation is being promoted through public education. Bruno Hénriquez, one of CUBASOLAR's founders, explains that because Cuba is an island with limited resources and because independence is a natural condition, Cubans know they need to turn off lights so that others will not have a blackout. He contrasts this feeling of unity and identity ("Somos Cubanos") with the way people in California acted in recent energy blackouts, suggesting the idea that the Californians had paid for the energy they were using contributed to a different perspective on energy conservation. The mindset to which Hénriquez calls attention is the sense of the common good of society as well as one's own interest. Rather than pursuing the maximization of self-interest in the pursuit of wealth, he says, ordinary Cubans recognize that their self-interest is inextricably linked with the good of the community and, indeed, the long-term viability of the natural resource base.

It is only recently that the public education campaign has shifted into high gear. The Electric Energy Conservation Program organized by Ministry of Basic Industry has as its primary objective the reduction of electricity use during the hours of peak demand. In 2003, a public education campaign on billboards began to be seen around the country. One billboard in particular depicts a smiling school-age child standing under a bright sun on a circle with trees and reads, "Conserve Energy, the Hope of the Future." In addition to linking conservation with children, who are highly valued in Cuban political culture, the sign also reminds Cubans of the benefits of trees and encourages them to plant them.

Schoolchildren learn about the various renewable resources of energy not only in the classroom but also firsthand, as many of the new renewable energy systems have been installed at school sites. The university and technical educational institutions are also incorporating renewable resource technology studies into their course offerings.

Working with CUBASOLAR, the Ministry of Higher Education has established the Technical University of Renewable Energy to train Cubans and students from elsewhere in Latin America (Barris and Madruga 2000: 20). Additionally, students training to be specialists in renewable energy can take classes at the José Antonio Echeverría Polytechnical Institute.

The public education campaign discusses the need to conserve energy in terms of simply using resources more wisely and utilizing new technologies such as compact fluorescent lights. Indeed these lights have been installed in hotels, restaurants, and even private homes. Further public education campaigns include energy saving tips including the placement of refrigerators to reduce energy, e.g. placement of the appliance in proximity to the wall for maximum air circulation, and basic maintenance, energy efficiency advice on aspects like checking refrigerator seals, as well as how to conserve energy more efficiently when utilizing air conditioning. For example, historically Cubans have kept offices and buildings very cool during the summer season, a shock to one's system to move from ninety degrees F plus heat and humidity levels to air conditioning in the sixties. There has been some progress in this area though more could occur in an informal survey of government offices in the summer of 2003. One new development is the utilization of a freon substitute in refrigerators. Cubans are using natural gas substitutes as well as experimenting with other substances to replace freon. Use of this substitute is indicated by labeling of refrigerators as energy-efficient.

Public education campaigns on energy conservation are not only limited to Cubans at home. Efforts are also underway to educate those involved in new building construction as well as the large endeavor of restoration occurring in Habana Viejo (old Havana). These efforts are aimed at teaching people about the importance of house siting, placement of trees for maximization of shade for natural cooling and lighting, as well as window types so as to best utilize solar heating and cooling.

According to Hénriquez, these public education conservation efforts already have begun to pay off. With the new technologies and conservation, the energy that was once used in a day now lasts fifteen to twenty days (Henriquez 2003). In 2003, Cuba reached a point of only utilizing the very small amount of oil it generated, according to Hénriquez (Henriquez 2003). Renewable forms of energy made up for the disparities between the nation's energy needs and the less available and more expensive petroleum, he explained.

GREEN MEDICINE

In addition to energy self-sufficiency, Cuba has been pursuing self-sufficiency in medicine, embracing the use of green medicines along with conventional allopathic medicine. This is in fact an expansion of development dating back to the 1940s, but the problems of the Special Period helped facilitate the reintroduction of green medicine into the repertoires of physicians and made it much more acceptable on a larger scale.

According to Mercedes García of the Pasture and Forage Research Institute, rural-to-urban migrants had brought with them plants and medicinal herbs and knowledge of their uses with them, and the lack of medicines during the Special Period simply allowed for greater adoption of these medicines which luckily had been well-researched (see Table 4.1) (García 2002: 212). One of the pioneers in this research, often called the "father of Cuban green medicine," was Dr. Juan Tomás Roig y Mesa, who wrote a number of books on the use of medicinal plants and herbs in the 1940s (García 2002: 212). His work laid the foundation for the 1973 creation of the Medicinal

Plants Experiment Station in Havana, which continued the research on green medicine and the dissemination of knowledge about their use for Cuban laypeople (García 2002: 213).

Fear of being cut off from the rest of the world by a military blockade during a conflict with the United States led the government to include plans for medical self-sufficiency (García 2002: 213). Given that they can be produced within the country, green medicines are more accessible to all and can even serve as a potential source of income and employment. To accomplish this pharmaceutical autarky, the military embraced green medicine, opening the Central Laboratory of Herbal Medicine at the Higher Institute of Military Medicine. The ministries of Science, Technology, and the Environment, Agriculture, and Public Health cooperated to facilitate its expansion, and permitted a streamlining of the processes for conducting clinical trials across the island assessing the medical effectiveness of plants and herbs. Special attention was focused upon the medicinal plants' usefulness to Cuba's elders (García 2002: 214, Acosta 1993: 20–25).[3]

Medicinal plants are derived from natural ingredients and, for Cubans, fall in line with the natural origin of humans (Acosta 1993: 20–21). Proponents of Cuba's green medicine see the shift as a "return to nature." The fact that the herbal medicines can be grown in flowerpots, patios, and kitchen gardens demonstrates how this alternative form of medicine can play a role in the island's effort toward self-sufficiency.

Cubans have also been crafting new medicines from natural products such as the widespread use of honey and bee products. ApiQueen, an enterprise of the Corporation of Tourism and International Business, produces products from honey, bee pollen, propolis, and royal jelly. According to an ApiQueen brochure, honey not only is used as a source of nutrition but also has germicidal and revitalizing properties. Bee pollen is said to have beneficial impact on the digestive system, and to be useful in preventing migraines, hypertension, and stress. Royal jelly considered high in protein, sugars, and antibiotics aids anemia and the endocrine and circulatory systems. Bee propolis, derived from bee secretions of plant resins, is said to have anti-inflammatory, antifungal, and antiparasitic properties.

In addition, Cubans are also experimenting with foreign forms of alternative medicine such as Bach Flower remedies and homeopathy from Germany and acupuncture from China. They also work with techniques from massage, Reiki, and hydrotherapy (e.g., sulphur and/or mud baths).

Alternative medicine is used alongside traditional Western medicine in Cuba's 169 municipalities. It has been part of the medical school curriculum since 1994, where it includes acupuncture, herbal medicine, and homeopathy (Beinfield 2001). Since 1995 medical students have been able to specialize in traditional and natural medicine, which adds four years of training to the six years of medical school training and three-year residency

Table 4.1. Reported Medicinal Uses of Familiar Plants

Illness or Therapeutic Property	*Useful plants*
Common cold	Aloe Garlic Eucalyptus Mint Plantago (*Plantago major*) French oregano (*Coleus amboinicus*) Sage Linden
Antirheumatic and anti-arthritic	Garlic Onion Cabbage Eucalyptus Pine Sunflower Chili pepper
Bleeding gums	Avocado
Heart tonic	Corn
Astringent	Strawberry Avocado Citrus Guava
Dermatitis	Calendula (flower) Eucalyptus Chamomile Oregano Guava
Prevention of boils	Garlic Coconut Okra
Psoriasis, vitiligo, skin cancer	Rue
Antispasmodic	Calendula Chamomile Passionflower Rue Tamarindo Cilantro Lemon grass
Bronchitis	Mango Onion Eucalyptus Cabbage

Table 4.1. (*continued*)

Illness or Therapeutic Property	*Useful plants*
Insomnia	Chamomile Passionflower
Sedative	Calendula Chamomile Passionflower
Antiviral, flu	Coconut Cabbage Papaya Citrus Tamarindo
Laryngitis	Onion Papaya Eucalyptus
Stomach ache/gastritis	Sour orange Anise Mint Fennel Lemon Plantago Ginger
Cough	Eucalyptus Ginger French oregano

Sources: Acosta (1993), Fuentes and Granda (1997), García in Funes et al. (2002).

in family medicine. The turn toward alternative medicine has also produced a new office of Natural and Traditional Medicine within the Cuban Ministry of Public Health. According to the Pan American Health Organization, nearly half of the Cuban population sought traditional and natural treatments in the mid-1990s (Kovaleski 1999: A10). The alternative medical sector witnessed a huge increase in the number of Cubans seeking care. Leoncio Padrón, director of traditional and alternative medicine in the Ministry of Public Health, reports that the number of patients treated with alternative medicine (acupuncture, homeopathy, and herbs) "more than doubled to about three million in 1998 from 1996" (Webster 1999).

Marta Pérez, director of natural and traditional medicine for the Ministry of Public Health in Havana Province, explained that the government promotes natural medicine because it is sustainable and cost-effective. "The Special Period has been a great teacher for Cuba," she said, "because in the midst of this difficult situation, we had to find a way to fight back" (Webster

1999). There was resistance to the addition of alternative techniques, she acknowledged, but the government insisted it was important to have treatment options and looked mainly for techniques that could be defended scientifically (Webster 1999). The adoption of green medicine reflects in part the response of Cubans themselves to these types of medicines, in other words, the medical community paid attention to what average Cubans were doing and thinking. In many ways, sustainable development depends on the participation and creativity of its human participants.

The Cuban experiences demonstrates that with the right mix of political and economic conditions as well as research and action by scientists, technicians, and citizens working for a better society, great progress can be made in the transition to a more sustainable society. Clearly a nation must focus on meeting the needs of its citizenry. As we saw in the earlier chapters on agriculture and food security, in this section we see Cuba's efforts toward moving to greater energy and medical self-sufficiency.

As scientists around the globe have already forecast, the world will run out of fossil fuels in the twenty-first century. Recognition of that fact necessitates preparation of renewable resources that allow for the economic development necessary to support human development as well as reduce harmful emissions that contribute to global warming. Indeed these are concerns shared by all the nations of the world. A nation, such as Cuba, with small domestic sources of petroleum, faces a greater urgency in developing alternatives to fossil fuels. If a developing nation of the Global South must obtain petroleum products from abroad then that nation must use precious foreign capital reserves and at the same time become vulnerable to the volatility in prices for fuel on the open market. As I have stressed earlier, self-sufficiency is not an end to trade, but in a sense, it is a national defense. Energy self-sufficiency by reliance on domestic renewable resources provides a cushion against the market volatility, which can have great economic impacts on a nation and impair its ability to provide the means of human and social development. Further, nations of the Global South often charge the dominant nations of the Industrialized North as controlling and even manipulating global energy markets. But as Bruno Henríquez quipped, "the sun is democratic—it is for everyone" (Henríquez 2003).

Energy self-sufficiency also provides a means of development that does not impair intergenerational equity and intergenerational equity, and reflects Cuban commitments in long-term sustainable development. While Cuba has made much progress in the areas of industrial and household energy self-sufficiency, work is only just beginning in the area of transportation. Cubans are in the process of research and utilizing renewable fuels, electric cars, and biodiesel (Henríquez 2003). Biodiesel fuel is created from biomass products, often leftover materials from agriculture or industry. The Cubans are also exploring solar energy for transport as seen recently in the creation of a solar-powered boat, and they are also exploring the most energy-efficient ways to produce hydrogen, which when burned simply produces water as a by-product.

Similar to alternatives to traditional petroleum-based fuels, the Cubans have been researching and utilizing alternatives to traditional Western allopathic medicine. Cubans, Europeans, and North Americans have set up research centers on alternative medicine as well as incorporated their use along with allopathic medicine at hospitals and doctors' clinics, but Cuba has gone a step further in adding alternative or green medicine as a focus of study in their world-famous medical education programs.

Here again we see efforts to create a more sustainable society in terms of protecting Cuba from the economic volatility of purchasing allopathic medicines on a world market, particularly in light of the nearly forty-five-year-old economic blockade by the United States. Some Cubans which also maintain specific techniques, such as with the use of herbal medicines, also has a lighter footprint on the earth than do other techniques, alternative and allopathic.

The transition in medicine was partly due to the realization of Cuba's medical community that ordinary Cubans already utilized these forms of medicines, along with the Western allopathic medical regimens used by most Cuban physicians. This aspect attests to the importance of participation in creating a sustainable society at all levels. (One cannot though discount the need to supplement and even substitute for medicines that simply are difficult or impossible to obtain given the blockade as the strongest driving force in the development of alternative medicines.)

NOTES

1. Programa de Desarrollo de las Fuentes Nacionales de Energía was hammered out by Raúl Castro, vice ministers from the various agencies within the central administration of the state, as well as the national energy commission, the economic planning commission, and the National Academy of Sciences. Luis Bérriz, "Las energies renovables en Cuba," in Enegía y tu, No. 22, CUBASOLAR. Dr. Bérriz is a senior investigator for CUBAENERGíA and president of CUBASOLAR.

2. González Couret is vice dean of the Architecture Faculty at the Facultad de Arquitectura, of the Instituto Superior Politécnico José Antonio Echeverría and a member of the board of directors of CUBASOLAR.

3. For more on Cuban medicinal plants, see Fuentes and Granda 1992.

5

Pearl of the Sea

Cuba, entre el flotante velo
allá te vemos brillar
como una perla del mar
entre los tintes del cielo —José Joaquín Palma[1]
(Cuba, between the floating veil, we see you shining like the pearl of the sea between the colors of the sky)

To those who have never visited Cuba, mention of the name may conjure up visions of bearded revolutionaries, fifties-era American automobiles, and legendary *Cohiba* cigars. But the island is—ideological debates aside—a place of dreams: stunningly beautiful, with sparkling white beaches and coastal waters from periwinkle to violet, mountains swathed in verdant forests, and coral reefs teeming with a kaleidoscope of living color.

Cuba is a beautiful island with a rich diversity that in a number of ways parallels Amazonia. Following the principle of biogeographical regionalism, because Cuba is an island, the larger the island, the higher the number of endemic species. The country has a number of species unique to the island at a rate ten times that of most of Europe, and the number of native species surpasses that of much of Siberia as well as likely Brazil's section of the Amazon. Cuba's coral reefs are only surpassed by those in Australia's Great Barrier Reef. It is also home to the greatest diversity of flora within the Antilles and the number of its endemic species of plants is only exceeded by the Cape region, Hawaii, and parts of Australia. But its plants and animals are endangered. Nearly one thousand species—of which 832 are found only in Cuba—are endangered. Undoubtedly, given this level of biological diversity and endemism, the environmental patrimony of the island must be protected.

Cuba is home to the tiniest hummingbird (*zunzuncito*) in the world—the 63 mm long bee hummingbird (*Mellisuga helenae*), the fierce *cocodrillo* Cuban crocodile (*Crocodylus rhombifer*), six species of spectacularly colored snails (*Polymita*), and the tiny Iberian frog (Rana iberica), to name but a few of the panoply of Cuba's endemic species (Gutiérrez Domech and Rivero Glean 2002: 78). According to the Cubans, Columbus is reported to have seen a manatee resting on rocks along the coast and mistaken it for a mermaid.

The manatee (*Trichechus manatus*), an endangered species in south Florida, is also found in the waters off the Cuban coast. Strict regulations on the manatees have helped bring them back from the brink of extinction twenty years ago in Cuba (Gutiérrez Domech and Rivero Glean 2002: 79). Another species believed to be on the brink of extinction, the ivory-billed woodpecker (*Campephilus principalis bairdii*) may still possibly live in the rainforests in the Toa River basin based on Cuban observations and the recent sighting of the ivory-billed in Arkansas. The national bird of Cuba is the tocororo with its feathers of blue, red, and white (*Priotelus temnurus*) the vision of which is said to evoke feelings of living in a fantasy world (García Montaña 2002).

While signature megafauna animal and bird species such as these can serve as rallying symbols for environmental protection, Cuba's biological resources are not limited to these creatures. As on any island, given the rules of adaptation and evolution, biological diversity is extensive. In terms of biodiversity, Cuba is viewed "as the most important country in the Caribbean" (Smith 1995: 13). The variety of soils, lithologies, and climate make possible a wide diversity of vegetation and ecosystems. "The endemism in the Cuban archipelago is 10 times greater than in the majority of European countries, and other new endemic species are continually being discovered. There are more endemic species in Cuba than in the unlimited Siberian expanses of the former Soviet Union and probably more than in the entire Amazon region of Brazil (Cuevas 1995: 24).

The Cuban archipelago, sometimes referred to as having the shape of a crocodile, has an area of 110,992 square kilometers and includes two main islands, Cuba and the Island of Youth (Isla Juventud) as well as nearly 4,200 cays or keys and smaller islands. It measures 1,200 kilometers in length from Cabo San Anotonio to the point of Maisí (Núñez Jiménez 1965: 25). The width of the archipelago is greatest between Taralogos (or Tarascos) Beach in Camagüey and Camarón Grande Point in southeast Granma Province, about 191 kilometers (Gutiérrez Domech and Rivero Glean 2002: 15). Its narrowest point, only thirty-one kilometers, lies between Del Rio Inlet and Majana Inlet, both in La Habana province (Gutiérrez Domech and Rivero Glean 2002: 15). "Seventy-five percent of the territory is made up of plains, 18 percent of mountains, and the remaining 4 percent of moist soil areas. The high level of biodiversity can also be appreciated in the various ecosystems and landscapes, which range from semi-deserts and arid mountains to tropical rain forests" (Cuevas 1995: 24).

The Cuban landscape is comprised of rainforest, savannahs, wetland, and semi-desert areas. The variety of the landscape results from what Cuba's best-known modern scientist, Antonio Núñez Jiménez, called its double insularity. "Cuba is twice over an island. In the first place, it is surrounded by sea and secondly, its insular shelf is separated from the shelves of the North and South American continents and from the rest of the Caribbean islands" (Núñez Jiménez quoted in Gutiérrez Domech and Rivero Glean 2002: 15).

The climate of Cuba is influenced by two powerful ocean currents, the north and south trade currents, which converge to create the Gulf Stream. The waters off the coast range from 26°C in the winter (79°F) to 29°C (84°F) in the summer. Temperatures (without humidity indexes which can substantially change the heat index and comfort levels for humans) vary from about 7.5°C to 21.5° C (46°F to 70°F) in January, the coldest month, to 30°C to 20°C (86°F to 68°F in late July to mid-August, generally the hottest part of the Cuban summer). Rainfall is abundant but not excessive except during hurricanes or tropical storms. The rainy season is generally in the summer months and the dry season falls in winter. Rainfall is higher in the mountainous areas, especially on the western side of the island (1,400–1,600 mm), than in the eastern plain areas (700–1,300 mm) (Del Risco 1999: 29).

Across the island, the most important rivers are the Cuyaguateje located in Pinar del Rio Province; the Cauto, the island's largest river, which runs unusually from east to west through Granma, Santiago del Cuba, and Holgúin; the Agambama, running through Villa Clara and Sancti Spíritus; the Almandares, traversing Havana Province; the Máximo in Camagüey; and the Toa, the nation's most powerful river, in Guantánamo. The depth of these rivers and therefore their suitability of navigation vary with the rainy and dry cycles. Most rivers are often quite shallow and not suitable for transport.

Cuba is home to a number of soil types: ferritic, ferrallitic, fersiallitic, brown soils, calcimorphic, vertisol, hydromorphic, hallomorphic, alluvial, and evolved (quarzitic sandy, calcareous sandy, and skeletal soils) (Gutiérrez Domech and Rivero Glean 2002: 79–81). The variety of soil types also contributes to the diversity of vegetation on the island. A defining feature of the Cuban environment is karst: limestone with fissures, dolines and other depressions, caves and sinks, and underground watercourses (Tell 1973 cited in Gutiérrez Domech and Rivero Glean 2002: 45). Over the ages, with the cycles of rain and drought, the erosion of the carbonate rocks has led to the karst buildup seen in the buttes called mogotes, and the island's caves and canyons (called *hoyos de montaña* or *hoyos de terreno*).

COASTAL AREAS AND WETLANDS

The Cuban archipelago has a number of bays and harbors: Havana Bay, San Pedro, Cienfuegos, Cochinos (Bay of Pigs), Nipe, Guantánamo, Santiago de Cuba, Nuevitas, Puerto Padre, and a multitude of smaller bays along the northern coast and archipelagos Sabana and Camagüey also known as Jardines del Rey. Cuba's coral reef is so extensive it is seen as second in the world only to Australia's Great Barrier Reef (Gutiérrez Domech and Rivero Glean 2002: 88). Other archipelagos include Jardines de la Reina, Los Canarreos, and Los Colorados. The areas range from having good to poor marine bottoms for navigation to sandy beaches to muddy mangrove coasts. These features will likely encourage tourism or discourage tourism, which in all likelihood will contribute to their preservation. Coral reefs and mangroves serve as natural barriers to hurricanes and tsunamis.

WETLAND AREAS

Wetlands cover about 8.25 percent of the island. Five areas of them have been nominated for the List of Wetlands of International Importance. In addition to the Zapata Swamp, which had already been set aside under the Ramsar convention of 1971, these are the Buenavista Bay (currently a national park, protected area, and UNESCO biosphere reserve), the Lanier (Isla de Juventud), the Ciego de Ávila, the Cauto Delta, and the Río Máximo-Cagüey (ENS 2002).[2] A variety of flora and fauna fall under the protection of the designation, among them greater flamingos, whistling ducks, manatees, and crocodiles. The Cauto Delta wetland, at 47,836 hectares, is viewed as the second most important in the country. A multitude of birds, 105 species, and some sixteen species of reptiles, including the anole de birama, iguana, and the American crocodile (*Crocodylus acutus*) are found there (Ramsar-Cauto 2002).

The Buenavista Bay wetland, 313,500 hectares, is composed of beach systems, dunes, salt flats, and lagoons (Ramsar-Buenavitsa 2002). In addition to important Cuban plants, such as *The Cactus Ripsalis cassutha* and *The Orchid Oncidium guttatun*, the wetland is home to the American crocodile and American pink flamingo, the manatee (*Trichecus manatus*), the Cuban sandhill crane, and the tocororo (*Priotelus temnurus*). Especially of note in the area are the cave-dwelling bats and two rare hutias (jutia conga and jutia caraballi).

The Lanier wetland is made up of a variety of biotopes including semi-deciduous forest, lagoon reefs, marine grassbeds, mangroves, and swamp grasslands and consists of 88,160 terrestrial and 38,040 marine hectares (Ramsar-Ciénaga de Lanier 2002). It contains a number of endemic protected species, including fifty-seven species of birds. The Cuban crocodile (*Crocodylis rhombifer*) is found only here and in the Zapata Swamp, along with the American crocodile and the endemic and endangered Cuban tree boa (*Epicratus anguilfer*).

The 226,875-hecatare Ciego de Ávila wetland includes two important lagoons and wildlife protection zones, particularly for migratory birds, including rare and endangered such as the Peregrine Falcon (Ramsar-Ciego de Ávila 2002). It is also home to the endemic and rare Wax Palm (*Copernicia fallense*).

The Rio Máximo-Cagüey wetland is a very fragile coastal marine ecosystem. It includes a number of isles located in the shallow water of the deltas of the two rivers, an area of over eight thousand terrestrial and thirteen thousand marine hectares (Ramsar-Rio Máximo-Cagüey 2002). It is reported to be the most important nesting site for the American pink flamingo in the Caribbean. The area also serves as home to large numbers of crocodiles and manatees. The Cagüey area is seen as the more significant of the two, because of its flora. Of special note are two local species, *Copernicia cowellii* (Arecaceae) and *Ginoria arborea* (Lythraceae). In addition to the flamingos, there are twnty-thre species of endemic birds, among them the tocororo, and *Xiphidiopicus percussus percussus* (Ramsar-Rio Máximo-Cagüey 2002).

Zapata Swamp, at 452,000 hectares, is the island's largest and most important karstic drainage, and is of particular biological significance. It is said to be the largest and best-preserved wetland in the Caribbean, and also is the largest area of tide pools on the island (Ramsar-Ciénaga de Zapata 2002). The government designated it a multiuse protected area in January 1995 and a UNESCO Biosphere Reserve in January 2000. It is of particular note because of its plant life and also serves as an important stopover for migratory birds as well as supporting a number of important Cuban endemic and endangered species such as the Zapata rail, the Cuban tree boa, the Cuban crocodile, and the Cuban sandhill crane. An important conservation effort for breeding and reintroduction of Cuban crocodiles is conducted at a nearby crocodile center founded in 1959.[3] The center breeds about eight hundred crocodiles and releases about two hundred one-year-old crocodiles per year. Some of the crocodiles bred will die of disease and some are sold for meat (roughly 2 percent). Leon Felipe Pérez Perera of CITMA estimates that seven thousand Cuban crocodiles remain in the wild (Pérez Perera 2003). Presumably selling a small proportion of crocodiles for meat will reduce the number hunted in the wild.

In addition to the Zapata Swamp and Buenavista Bay, there are four biosphere reserves. The Sierra del Rosario reserve, declared in 1985, was the nation's first. The 26,686-hectare reserve is made up of evergreen tropical, semi-deciduous forest, mogotes, and coniferous forest with pine. The reserve protects the tocororo, trogons, sparrow hawks, as well as hutias, and a number of bat species. The Guanahacabibes Peninsula reserve, 119,189 hectares, is made up of semi-deciduous and evergreen forests, marshes, and coastal vegetation and contains such important endemic birds as the Cuban bee hummingbird (*Mellisuga helenae*), and other animals such as lizards, hutias, and iguanas (UNESCO 2003, Herrera et al. 1993). The Cuchillas del Toa reserve, established in 1987, covers 208,305 hectares, of which roughly six thousand are marine, and is one of the principal centers of biodiversity on the island, with mountain rainforests, cloud forests, xerophytic matorral areas, coral reefs, and mangroves (UNESCO-Cuchillas del Toa). It also includes a significant cave and houses a multitude of flora and fauna species. There are 928 endemic species such as the podocarpus and dracaena in the plant community and the caguarero sparrow hawk (*Chondrohierax wilsoni*), the almique (*Solenodon cubanus*), and the royal woodpecker (*Campephilus principalis*) (UNESCO-Cuchillas del Toa). The reserve is also home to the colorful Cuban painted tree snail (*Polymita*) (Herrera et al. 1993: 15). Three new species of Anolis lizards have been found there along with the Iberian frog (*Eleutherodactilus Iberia*), which is considered the smallest frog in the world (UNEP Alejandro de Humboldt). The extremely rare Cuban kite (*Chondrohierax wilsoni)* is also found in the area. The endemic and endangered royal woodpecker and perhaps even the ivory-billed woodpecker (*Campephilus principalis bairdii*) have been reported.[4] The Cuchillas del Toa includes the Alexander von Humboldt National Park, another World Heritage site. The Baconao reserve comprises 32,400 hectares and includes mountain rainforests, cloud forest, evergreen sub-mountain forest, coastal areas, and even cacti and cave areas (UNESCO Baconao). It supports a wide variety of plants, some 1,800 species, as well as a number of threatened endemic bat, spider, and insect species (UNESCO Baconao).

According to Bohrhidi and Muñiz, some 994 species belonging to 381 genera and 105 families are endangered today, and of these 832 species are exclusive to Cuba (as cited in Gutiérrez Domech and Rivero Glean 2002:6 2). Species density, particularly in an island setting, makes Cuba's biological diversity all the more significant: "The number of species per kilometer, in Cuba is 6 times greater than in California, 10 times greater than in Texas, 7 times greater than in Spain, 9 times greater than in Chile, 22 times greater than in Nigeria, 36 times greater than in India and China, and 42 times greater than in Australia," according to the Cuban scientist Jorge Ramón Cuevas (Cuevas 1995: 25). "The flora of vascular plants consists of 6,700 species, with an endemism of 51.4 percent. This represents the greatest diversity of flora within the Antilles, which has 70 genera and more than 3,000 endemic lower taxa" (Cuevas 1995: 25). The Cuban flora has one of the highest levels of endemism in the world, surpassed only by Africa's Cape region, Hawaii, and some parts of Australia (Table 5.1) (Cuevas 1995: 25).

Table 5.1. Current Knowledge of Cuban Fauna

Kingdoms and Subdivisions	*Number of Known Species*	*% Known*	*% Endemic*
INSECTA	6348	60	-
Orthoptera	250	40-70	70
Dermoptera	19	40-70	-
Isoptera	22	90	-
Odonata	85	80	12.5
Hemiptera	549	40-70	-
Homoptera	650	40-70	-
Lepidoptera	800	40-70	10
Diptera	929	40-70	30
Coleoptera	2170	40-70	50
Hymenoptera	910	40–70	25
ARACHNIDA	1700	70	75
Aranea	718	70-80	84
Acarina	606	40-70	70
MOLLUSCA (Terrestrial)	1700	70	87.5
CRUSTACEA (Terrestrial and freshwater)	60	40-70	30
NEMATODA, TREMATODA, AND CESTODA	600	90	35
(Zooparasites)		15	80
REPTILIA	105	60-80	80.9
AMPHIBIA	42	40–70	55
PISCES (freshwater)	58	40–70	55
AVES	330	100	6.6
MAMMALIA	88	100	40

Source: Cuevas 1995: 26.

PLEAS FOR PROTECTION

There have been pleas for the protection of Cuba's rich biological diversity from inside and outside of the island, by proponents and even critics of the government. These calls reflect concern over the loss of global biological diversity and point to similar problems facing nations trying to balance human needs with environmental protection. Edouard Saouma points to the need for humans to begin to value natural resources as valuable even though they are not traded on the open market. "A critical deficiency of the international economic system has been its ability to assign a value of exchange to biodiversity or other environmental elements," explained Saouma, former director of the Food and Agricultural Organization of United Nations. He said we face the challenge of "getting the formula right for incorporating the cost of preservation into the cost of production" so that we may fulfill our responsibilities to future generations and to halt the attacks on biodiversity (Saouma 1993 quoted in Cuevas 1995: 30).

The Cuban scientist Jorge Rámon Cuevas advised us to take a long-term perspective about humans' impact on the environment. "The cost of preserving biodiversity is much less than the punishment for permitting its degradation. Once this patrimony is lost, it will not be able to be recovered. Therein lies the challenge for each and everyone one of us," said Cuevas (Cuevas 1995: 30). He called on governments, scientists, politicians, businesspeople, and average citizens to take on these responsibilities and work together to solve these problems. "Society must recognize that biological diversity must be protected and simultaneously utilized. Conserving biodiversity means taking protective measured for genes, species, habitats and ecosystems," he explained (Cuevas 1995: 30).

"Cuba, which comprises more than one-third of the land area in the Caribbean, boasts nearly four times as many plant species as Jamaica and almost 12 times as many as Puerto Rico," according to Michael Smith, director of the Caribbean Biodiversity Program and the U.S.-Cuban Scientific Exchange at the Center for Marine Conservation (Smith quoted in Simpson 1999). Smith maintains for this reason, Cuba is the most important island in the New World.

CONCLUSION

Clearly, given Cuba's vast biological resources, national efforts for environmental protection should be supported by the international community. As we will see in following chapters, Cuba has embarked upon far-reaching and ambitious legal protection for its natural resources as well as society-wide environmental education efforts, while Cuban intellectuals search for ways to craft a new national environmental ethic and model for sustainable development.

NOTES

1. Palma quoted in Rivero Glean, 1999: 46.
2. The 1971 Convention on Wetlands of International Importance was signed in Ramsar, Iran, to improve, conserve, and manage environmentally sensitive internationally significant wetlands, setting up a global framework to fund and monitor the wetlands and acquire conservation and protection agreements from its signators. Cuba signed the Ramsar Convention in 1971 and it entered into force in 1975.
3. Field notes, June 2003, Ciénaga de Zapata.
4. The royal woodpecker is known to reside there; other reports on ivory-billed woodpeckers are often repeated but have not been confirmed. Perhaps the reports are not so unlikely given the recent sighting of the ivory-billed woodpecker in Arkansas. For more on birds overall in Cuba, see Florentino García Moñtana's books on Cuban birds, *Las Aves de Cuba: Especíes endemicas*, 2002 re-issue, and *Las Aves de Cuba: Subspecies enémicas*, 1998, Editorial Gente Nueva, Havana, Cuba.

6

Codifying Environmental Protection

You cannot protect nature only by law. If you can convince ***people****, then they will do it freely.* —Jorge Rámon Cuevas

As we have seen, Cuba is an island rich with the treasures of sugar-sand beaches and verdant forests and home to a multitude of endemic species, the number of which some experts compare with the Amazon rainforest. But Cuba is also home to a population of 11.2 million and the country faces the balance of protecting both this magnificent biodiversity of non-human life with the very real needs of its citizens. Given Cuba's commitment to social justice and equity, the balancing act becomes even more challenging. Crafting a strong legal foundation for the protection of natural resources while ensuring the satisfaction of basic human needs was a necessary and important step for the island on its road to a more sustainable society.

Cuba's environmental framework is ambitious and far reaching. The law embodies important principles set forth by the Brundtland Commission and the signatories of Agenda 21, the Rio Declarations. Contained within the law is the commitment to social equity, equity between current and future generations, the precautionary principle, biodiversity, environmental education, control of environmental degradation, and balancing human needs with those of protecting nature. Two of the most important aspects of the law include, first, the environmental impact review process that necessarily balances economic development *for* human needs with those of the protection of nature and, second, the promulgation of the precautionary principle with a heavy emphasis and reliance upon Cuba's social wealth of scientific experts and technicians.

In this chapter, I trace the creation of Cuba's environmental law, through a number of milestones, as well as how the current law developed within the changing political contexts of Cuban realities, including the economic crisis of the Special Period. This crisis led to an opening of the country's economy to the outside world and necessitated a foreign investment law. Further, the island recognized that if the nation's patrimony were to be protected while economic development expanded, new regulations had to be crafted to manage this development. Finally, as Cuba signed onto international treaties and agreements, the island would have to update and create new policies to adhere to the precepts in these international pacts. These actions set forth an important legal step in the creation of a more sustainable society.

THE EARLY YEARS

During Cuba's colonial period (1492 to 1898), the government focused on regulating mines, waters, hunting, seaports, mountains, and soil (González and García 1998: 54).[1] After the "North American intervention" period (1898 to 1902) and under the republic era (1902 to 1958), five new areas were added to the list of regulations that had an impact on the environment: taxes, sanitation, work, public health, animal health, and tourism (González and García 1998: 54). Many scholars place the starting date for specific legislation with regard to the environment in the 1930s, when the island's first national park was created, El Cristal (González and García 1998: 43). A number of important decrees were issued regarding the protection of certain ecosystems in Cuba, among them establishing a flamingo reserve in Camagüey (Decree No. 803), certain mountain and forest preserves, and regulating use of the Zapata wetlands, but most observers consider these measures to have been ineffective (González and García 1998: 44–46). González and García report that many of the earliest decrees on environmental protection have since been countermanded, in any case they were never applied to the large latifundias (González and García 1998: 45). Even critics of Cuba's revolutionary government characterize the environmental law of the colonial period as "largely ignored and rarely enforced" (Díaz-Briquets and Pérez-López 2000).

Early accounts tell of a verdant-forested island in pre-Columbian times:

> And he [Columbus] climbed to a small mount to discover some portion of land and he was not able to see anything because of the large woods, that were very fresh and full of fragrance. . . . Then, the young shipmen gave voices saying that they saw pines. I looked to the sierra and saw them so large and marvelous that I could not estimate their height and straight shapes that were as

> stout and thin as spindles. Then we realized we could build ships and an infinite number of decks and sheatings, and masts for the largest ships of Spain. I saw oaks, madroños and a fine river, and riggings to make water-powered saws. . . . Then we found a lateen rig and a section for the mizzlemast for the caravel La Niña. . . . And all the sierra were full of pines and all over the place there were diverse and beautiful timberlands. . . . And in that same fashion there were valleys, just like the mountains, full of tall and fresh trees that were a glory to see and pine forests that seemed to be plentiful.[2]

Del Risco points to this passage as indicative of what was to come for Cuba: "It was the dangerous admiration of a pair of pragmatic eyes that saw wooden planks instead of trees. But for the time being, Cuban forests did not suffer the massive assault of Spanish axes. Only after 1511, with the beginning of the Spanish conquest of Cuba, did the presence of the Spaniards begin to be important enough for the vital conformation of the island. . . . Fortunately for the forests, the newcomers were quite busy exterminating the native population and searching for gold" (Del Risco 1999: 18–19).

After the indigenous peoples were decimated and gold acquired, the Spaniards set to transforming the Cuban landscape into a large latifundia for raising cattle. The forests were cleared so that cattle could be run, and to make way for small subsistence agricultural plots and construction needs, and rare woods were supplied to the Spanish crown. Among the trees targeted were mahogany (*Swietenia mahagoni*), cedars (*Cedrela odorata*), and sabicu (*Lysiloma sabicu*) (Del Risco 1999: 20). Del Risco argues this process demonstrates the Spanish (and presumably the developing Cuban) mindset with regard to forests. As in Spain, he explains, forests were seen as common property that could be utilized only by high-ranking government officials. The Spanish crown saw the forests as a strategic reserve for the construction of ships and other projects, and limits were placed on the amount of wood that citizens could harvest even on their own property. Del Risco argues that mindset, combined with slow population growth, actually kept initial forest destruction levels down (Del Risco 1999: 20–21).

It was only a half-century later that Cuban forests faced what Del Risco calls their mortal enemy, sugarcane. Introduced to the island in the early days of Spanish colonization by Governor Diego Velázquez, sugarcane got its start on a large scale in 1590 (Del Risco 1999: 21). Trees were felled for lumber to build the sugar mills, for land to grow the cane, and for fuel to run the mills. Matanzas and Havana Provinces suffered the brunt of the assault leading to irreversible changes in their ecosystems (Del Risco 1999: 23). It is estimated that between 1775 and 1827 nearly 1.7 million hectares of forested land was lost, with some 60 percent of that area being used for sugarcane and other agricultural uses (Del Risco 1999: 23). In 1827, there were some 7,507,814 hectares of forests, but between 1828 and 1900 approximately three million hectares of forests were lost (Matos 1972 as cited in Del Risco 1999: 23). Forests began to be viewed as the locus of barbarism that needed to be brought under control to permit the continuation of progress.

As the sugarcane industry expanded and Cuba developed, the forests declined. Low land prices, cheap labor for sugarcane production, and U.S. investments through the 1903 U.S.-Cuba Trade Agreement sealed the fate of many forests. Camagüey and Oriente Provinces were especially hard hit in this period, with some 1.3 million hectares of forested land lost. In the early 1900s, only the highest forests remained intact, "the Guanahacabibes and Zapata peninsulas, the Isle of Pines (today's Isle of Youth) and plains to the south of Ciego de Ávila and Camagüey, and the central-western region of eastern Cuba (the Cauto Valley)" (Del Risco 1999). By 1958, only 16 percent of the island, some 1.8 million hectares, remained forested (Del Risco 1999: 21).

After the 1959 revolution, there were efforts to modify forest policy, and an ambitious reforestation plan was crafted. However, other priorities such as infrastructure construction (dams, roads) and the expansion of sugarcane production, blocked these efforts. In particular, following the large 1970 sugarcane harvest, the forested areas in Cuba reached their lowest level in its history, at only 15 percent. In the final decades of the twentieth century, tree felling continued in a moderate way, mainly in the most affected and impoverished forests, but silvicultural activities, including reforestation, managed to achieve modest results. There was a slight increase in forested areas to 18.7 percent in 1990, and in 2002, the figure was 22.8 (Cuevas 1995: 25, CITMA 2003). More recently, José A. Diaz, deputy head of the Ministry of Science, Technology, and the Environment, reports the forested area of the island, 23.6 percent in 2004, has helped reduce air contaminants by 3.6 percent (Riera 2005).

MILESTONES IN ENVIRONMENTAL LAW

Cubans identify six milestones of environmental protection leading up to the passage of the Framework Law on the Environment (Law Number 81) of 1997: (1) Article 27 of the Cuban Constitution of 1976, amended in 1992, which called for protection of the environment and, after the signing of the Rio Convention on Biodiversity in 1992, linked it with the idea of sustainable social and economic development; (2) Law 1323 establishing the National Commission for the Protection of the Environment and Conservation of Natural Resources (COMARNA) in 1976; (3) Law 33 on the protection of the environment and the rational use of resources on January 10, 1981; (4) Decree-Law 118 (January 18, 1990), establishing the National

System for the Protection of the Environment and charging the National Commission for the Protection of the Environment and Rational Use of Resources with responsibility for developing environmental policies at the national level and overseeing compliance with environmental regulations; (5) the establishment of the National Environment and Development Program (1993), in line with the acceptance of the United Nations' Agenda 21; and (6) the creation of the Ministry of Science, Technology, and the Environment in 1994 (CITMA 1999: 11, González and García 1998: 55–60).[3] I would add to these the collapse of the CMEA, the passage of the Foreign Investment Law in 1995, current establishment of the environmental impact review process by Resolution 168, and the adoption of the National Strategy on Environmental Education in 1997.

In the early 1970s, Cuba began an intense accelerated program of industrial development aimed at reducing the difference in living conditions between rural and urban locations and slowing rural-to-urban migration (Gonzalez y Garcia 1998: 51). With the increase in industrialization came the realization that environmental regulation was needed and COMARNA was created in 1976. Law 33, which followed, was an attempt to elaborate and give meaning to the protection of the nation's environment and natural resource base set forth in Article 27 of the 1976 Constitution:

> The state protects the environment and natural resources of the country. It recognized its close link to sustainable social and economic development to make more rational human life and to assure the survival, the welfare, and the safety of future and current generations. It is incumbent upon the [responsible] to implement this policy. It is duty of the citizens to contribute to the protection of the water, the atmosphere, the conservation of the soil, plants and animals, and of all the [future] potential use of nature (Cited in EAN 1999: 17).

Cubans view Law 33 as an early normative expression of environmental policy and note with pride that it was one of the first in Latin America and the Caribbean (Gonzalez y Garcia 1998: 52).

The law outlines the responsibilities of the state and spells out the requirements for the protection of natural resources and their rational use, including water (arts. 22–38), soil (arts. 39–47), mineral resources (arts. 48–60), ocean resources (arts. 61–75), flora and fauna (arts. 76–86), the atmosphere (arts. 87–93), agricultural resources (arts. 94–100), human settlements (arts. 101–107), and landscaping and tourism (arts. 108–116) (Ley No. 33 1981). It begins by establishing its constitutional basis in the 1976 Constitution: "The environment constitutes a fundamental good of the society, whose protection and conservation is of vital importance for our State" (Preamble). Law 33 also acknowledges its socialist heritage: "In socialistic society, man constitutes the highest good, and the social character of property facilitates the decision-making process that guarantees the integral protection of the environment and the rational use of the natural resources, as opposed to the capitalistic systems where the interests of private property contradict those held by the larger society" (Preamble). It points to the importance of solving the problems of environmental deterioration that the country was left with following the 1959 revolution and the damage from economic development and ignorance about the use and care of nature in the post-revolutionary period. The preamble continues with a statement regarding the necessity of protection and rational use of natural resources for the improvement of life for current and future generations in the economic and social development plans of the Cuban state.

In its General Provisions, however, the focus is on "conservation, protection, improvement, and transformation of the environment and the rational use of the natural resources, in accordance with the integral development policies of the country, in order to optimally use the productive potential of the nation" (art. 1) (Law No. 33 1981: Chapter 1).

Here, the language is that of socialism, "The relationships to the environment are based in the principles of the property, socialized means of production and planned social economic development and are established by virtue of the necessary harmony of the dialectical unity of man and nature and of the balance achieved between the demands of the population and the possibilities of development of the natural resources in agreement with its potentiality" (art.6) (Law No. 33 1981: Chapter 1).

The law specifies conservation measures for the prevention of pollution and environmental damage, and environmental rehabilitation when possible (art. 7). It stipulates that investment projects approved by state agencies must include measures to protect the environment and the rational use of the natural resources (art. 11), and that producers of pollution and waste must provide for its treatment and disposal (art. 15).

The second chapter focuses on specifics relating to protection of natural resources such as plants and animals, water, marine, and mineral resources, and impact on the environment through construction, landscaping, and tourism. Among the broad guidelines is a brief mention of lower levels of consumption and possible reuse of water (art. 28).

There are moments when the law appears to anticipate the tension between economic and social development, and then protection of the environment that comes to the fore. For example, Article 37 reads, "The operation of the hydrological resources is governed by the norms established for the preservation of the aquatic life with economic aims or for the conservation of the ecological balance."

At other points, it suggests that economic and social development will always trump protection of the environment: Article 50 says that "mineral resources must be used in the most rational form possible to satisfy the needs of industrial, energy, constructive, and agricultural activities or the national interest."

Chapter 3 establishes the National System of Environmental Protection and Rational Use of Resources to facilitate the "harmonious participation" of government agencies and political, social, and mass organizations and the Cuban citizenry in environmental protection. The National System is charged with coordinating and implementing environmental protection and the rational use of natural resources to satisfy the needs of society; proposing and implementing the scientific norms and measures that guarantee the conservation of the environment; considering scientifically grounded environmental factors in social and economic development and eliminating or reducing the potential for environmental damage; and developing contingency plans for natural disasters as they relate to natural resources and the environment (art. 122). It also directs the system's Council of Ministers to craft regulations and laws for the protection of the environment and the rational use of resources (art. 124). Chapter 4 deals with enforcement and penalties for noncompliance of the law. It gives authorities the power to halt actions in violation of law and require remediation of harm to the environmental or natural resource base (arts. 127 and 128, Law No. 33 1981: Chapter 4).

Cubans soon identified a number of problems with this earlier environmental law. Roberto Acosta, an expert in oil and hydrocarbon pollution who served on COMARNA and the National Environmental Commission, pointed in particular to "a contradiction in the former management structure of the environmental policy in the country, in which certain ministries were administrators of environmental matters of the same resource that they exploited in order to fulfill their productive objectives. As a consequence of this situation, these administrators played the roles of 'judge' and 'party to the action' simultaneously, which led, on occasion, to faulty decisions and to little enforcement of the conditions established in the environmental evaluations that had been developed" (Acosta 1995: 65).[4] This difficulty was echoed by others, and some further argued that ambiguous lines of authority among the agencies often led to a stalemate in environmental disagreements (Everson 1998).[5] Indeed, the ambiguity exists in the law itself: "The protection of the environment and the rational use of the natural resources is responsibility of the State, the society, and the Individual, those which have the obligation of maintaining them in optimum conditions, in order to make possible the life in an environment adapted for the full development of its activities" (art. 4). The notion of protecting the environment so that development can occur has been central to the debates over sustainable development both in the Global South and the North.

Acosta goes on to call attention to the need to bolster state environmental inspection to guarantee legal compliance, clarify procedures for environmental impact assessments and implementation of recommendations that emanate from them, and to strengthen the relationship between environmental measures and national science" (Acosta 1995: 65). Acosta says proof of the nation's commitment to correcting past problems is seen in the three-fold increase in specialists charged with overseeing environment inspection even in a time of economic difficulty (Acosta 1995: 66).

Other Cubans, such as the president of the National Assembly, Ricardo Alarcón, and environmental ministry officials pointed out that COMARNA was unable to develop regulations specific enough to deal with the environmental issues it was required to oversee or to enforce any such regulations (Rey June 2003, Alarcon 1997, Everson 1998: footnote 138). These observers called attention to the fact that while decree-laws had been issued regulating noncommercial fishing, swine, plant safety, livestock, seed quality, beekeeping, soil conservation, forest and fauna protection, water resources, and fishing, marine resources, the atmosphere, human habitations, and tourist and scenic resources remained unregulated (*Alarcón* 1999: 21–22, Rey 1999: 12).

Additionally, the law did not cover the loss of biological diversity, the depletion of the ozone layer, global warming, and the international movement of hazardous wastes.

It became clear that a new law was called for to contend with the new political and economic realities of the Special Period. Confronting the acute lack of petroleum products needed for transportation, industry, and agriculture, as well as foodstuffs, medicine, and consumables, Cubans turned to their natural resource base in an effort to replace those products.

"The acute changes at the national level, generated by the economic crisis, reached a critical point at that time, and it was clear the crisis was rebounding in the environment . . . with an increase in the illegal cutting of trees and illegal hunting . . . but paradoxically it caused positive effects, encouraging a search for more sustainable alternatives, such a bio-pesticides and bioremediation, a healthy solution for a country like ours that had reached the point of having an excessive consumption of chemical products in agriculture," said Orlando Rey Santos, director of Environmental Policy Directorate, CITMA (Santos 1999: 12).

As the island opened itself up to foreign investors, mining and tourism were recognized as significant areas of foreign investment that could have a negative impact on the environment. The investors came from nations with high environmental standards and complicated legal regimens for overseeing the environment. Looking to see where they stood with regard to Cuban law, they found an aged law without concrete demands or assignment of responsibilities and lacking many of the basic provisions in this sphere and a system of outdated technical norms (Rey 1999: 12–13). Environmental

Ministry Official Orlando Rey Santos stressed that foreign investors did not dictate environmental legislation to the Cubans (Rey, Havana 2003). But as the economic crisis deepened, Cuba took steps to expand foreign investment and put in place a number of reforms in an effort to rebuild its economy. Law 77 on foreign investment, passed in 1995, established the legal framework for investment by foreign corporations in a multitude of sectors within Cuba, other than health and educational services and some sectors of defense (art. 10) (Law No. 77 1995: Chapter 1).

Under the law, the state is to "promote and encourage foreign investment . . . to carry out profitable activities which contribute to the country's economic capacity and sustainable development, on the basis of respect for the country's sovereignty and independence, and the *protection and rational use of natural resources*" (art. 1, Law No. 77 1995: Chapter 1). Foreign investors are invited to participate in Cuba's economy in one of three ways: in joint ventures between national investors and foreign investors, creating a Cuban commercial company; through international economic association contracts, and as commercial entities without the involvement of any national investor (art. 2, Law No. 77 1995: Chapter 1).

The new law requires that environmental conservation and the rational use of natural resources be carefully undertaken and establishes clear procedure for the evaluation of projects in terms of their impacts on the environment and natural resource base (art. 54). In the event of damage to the environment, the perpetrator is "obliged to reestablish the previous environmental situation, repair the material damage, and indemnify the injured parties" (art. 56).

The Ministry of Science, Technology, and the Environment issued regulations concerning environmental impact assessment on September 15, 1995, as Resolution 168. An environmental impact study in the language of the 1997 Framework Law is a "detailed description of the characteristics of a planned project or activity, including a description of its technology, which is submitted for approval through a process of environmental impact assessment. Well-founded information must be provided for the prediction, identification and interpretation of the environmental impacts of the project. Both the actions that will be taken to prevent or minimize any adverse effects on the environment as well as the monitoring program that will be adopted must be described" in the assessment (art. 8, Law No. 81 1997: Chapter 2). Environmental impact assessment is described as "a process whose object is to avoid or mitigate undesirable environmental effects resulting from plans, programs, projects, or activities through a prior estimation of the environmental changes caused by such projects or activities and, in appropriate cases, the denial of permits necessary for the execution of those acts or their approval with specific conditions." It includes detailed information about the monitoring and control system to ensure compliance and about the necessary mitigation measures that should be considered (art. 8, Law No. 81 1997: Chapter 2).[6]

An environmental license must be received from the Ministry of Science, Technology, and the Environment for any project that could have "significant environment effects" or in situations where a specific control is required to "conform with the requirements of existing environmental laws" (art. 24, Law No. 81 1997: Chapter 3). A license is obtained by following specific guidelines laid out by the Ministry and the Office of Environmental Inspections as well as paying a fee. Simply obtaining a license does not protect a project from any administrative, civil, or criminal liabilities if it injures the environment or an individual (art. 25, Law No. 81 1997: Chapter 3). Finally, the ministry can "temporarily or permanently suspend" a project, program, or activity that does not have a license or a license that is found to be in violation with the requirements and controls of the license.

Projects and activities requiring the environmental impact study (EIS) process include

> dams and related water projects; metallurgical, chemical or petrochemical plants; hazardous waste facilities and activities; mining activities; electrical plants; railroad facilities, airports, and harbors; refineries; tourism facilities, especially ones near the coast; large housing developments; industrial parks; agricultural, forestry, aquaculture or mariculture uses that could have great impact on endemic species, soils, or ecosystems; sewage systems; hydrocarbon extraction facilities; landfills; cemeteries and crematories; sugarcane, foodstuff, dairy, beverage, foundry, paper, cement, or auto industries; activities not included in the protected area's management plan; and any other activity that might significantly alter ecosystems, their composition or balance, or affect public access to natural resources and the environment in general" (art. 27, Chapter IV, Law of the Environment, 1997).

The law also stipulates that the parties applying for the environmental license will bear the costs not only of the license but the environmental impact study as well.

The Ministry of Science, Technology, and the Environment was set up on November 28, 1994, by Agreement No. 2823, by the Executive Committee of the Council of Ministers. Another agreement, No. 3264, passed February 19, 1998, added new responsibilities and clarified those previously identified.

According to the agreements, the functions of the ministry with regard to the environment include the following: (1) fashioning a national science and technological strategy in conjunction with the scientific community and other entities involved in technological change, establishing the objectives, priorities, and programs of that strategy, and, once approved, directing its implementation; (2) directing and coordinating with the scientific community the creation, execution, and evaluation of the national program of scientific and technological innovation and establishing the general guidelines for the creation of scientific programs; (3) participating in, evaluating, and controlling, in coordination with the relevant organizations, the development of the projects under the plan and establishing the budget for the scientific and technical activities in accordance with the national priorities; (4) directing, coordinating, and controlling the integration of scientific, technological, and productive factors throughout the scientific community; (5) facilitating the relationships between the various governmental and nongovernmental bodies within the scientific and technical communities; (6) directing and controlling the execution of the policies guaranteeing environmental protection and the rational use of resources in the sustainable development of the country, proposing and establishing the necessary national strategies for specific natural resource and biodiversity protection, and drafting and controlling the execution of programs for enhanced environmental control and the management of the agro-industrial waste and cleaner industrial production; (7) overseeing the implementation of the regulations established for the protection, conservation, and rational use of resources, resolving disputes among the organizations and agencies, in these areas, and approving the environmental impact assessments; (8) monitoring atmospheric pollution and climate conditions; (9) managing the country's biological reserves and protected areas; and (10) managing the strategies and programs for environmental education, both formal and informal.

In June 1997, the National Assembly approved an ambitious and far-reaching new environmental law, Law 81, which seeks to create a broad framework for environmental protection and management.

Assembly president Ricardo Alarcón, in a formal written statement, discussed the need for this new, more comprehensive law:

> Whereas Law 33 of January 10, 1981, Environmental Protection and Rational Use of Natural Resources, represents an early and important normative expression of the principles of Cuban environmental policy which established the basis for the development of a national legal framework in this sphere . . . the current conditions of economic and social development demand a legal framework more in line with the current situation. Law 33 has been surpassed in large part by the recent advances in environmental matters, both in the national and in the international realm. Accordingly, the aforementioned law must be replaced by a legal instrument that will adequately reflect the requirements of environmental protection and the attainment of sustainable development.

In the first chapter of the law, Article 3 sets forth the overall responsibilities of the government and the citizens toward Cuba's environment.

The State, the citizens, and the society in general have a duty to protect the environment through

a. conservation and rational use;
b. systematic struggle against the causes of its deterioration;
c. corresponding actions to rehabilitate it;
d. constant increase in citizens' knowledge of the interrelationship among human beings, nature, and society;
e. reduction and elimination of patterns of production and consumption that are environmentally unsustainable; and
f. promotion of demographic policies appropriate to local conditions (Article 3, Title I, Chapter I, Law of the Environment, 1997).

Under the new law, protection of the environment is to be based on use of natural resources in a rational manner to prevent negative environmental impact, environmental protection is to become part of development program planning at all levels, responsibility for environmental protection is to be both national and local levels, and the protection of natural resources is to be based on relevant research.

Interestingly, "In the case of a threat of major or irreversible harm to the environment, the lack of absolute scientific certainty shall not be used to justify a failure to adopt preventative measures" (art. 4-d, Law No. 81 1997: Chapter 1). This provision is an expression of the precautionary principle, established by the Rio Summit in 1992: when "in order to protect the environment, the precautionary approach shall be widely applied by States according to their capabilities. Where there are threats of serious or irreversible damage, lack of full scientific certainty shall not be used as a reason for postponing cost-effective measures to prevent environmental degradation." The precautionary principle places the burden of proof not on the public but on the entities proposing the project.

The 1997 framework law establishes a number of the instruments of environmental policy and management, including the National Environmental Strategy, the National Environmental and Development Program, environmental land use planning, environmental licensing, environmental impact assessment, the environmental information system, the environmental inspection system, environmental education, scientific research, technological innovation, economic regulation, the national environmental fund, and the relevant enforcement measures (Law No. 81 1997: Title 3).

According to Oliver Houck, professor of Law and director of the Environmental Law Program at the Tulane University School of Law, the framework law

> is more ambitious in its goals and its details than any comparable legislation in the United States or Western Europe because, among other reasons, it was

> started relatively *de novo*. Its 163 separate articles embrace what would be, in the United States and the European Union, separate programs in air, water, waste, noise, toxic substances, historic preservation, biological diversity, national parks, forests, and wildlife refuges, coastal zone management, education, research and technology, and environmental impact assessment and planning. Additional chapters deal with the authorities of sister ministries in, for example, energy and agriculture. . . . This is a sweeping initiative. It is hard to think of a significant environmental issue omitted—which makes the task ahead, the implementation of these provisions, all the more daunting.[7]

The Law of the Environment also empowers CITMA in ways new to Cuba. The first is to elevate to the status of law an environmental impact review process managed by CITMA itself that will, on its face at least, include both individual projects and, as in for example the United States, the plans and programs of other agencies. To the extent that CITMA approves the environmental impact review process for all agencies and the adequacy of the reviews themselves, the Ministry's authority is, obviously, considerable. The second authority, also elevated to law, is an environmental license required by CITMA for specific proposals affecting the environment, in addition to whatever approvals are required by other agencies (e.g., for facility siting by the Ministry of Planning, and for pollution permits by the Institute of Water Resources). Environmental impact assessment in Cuba, therefore, as in many developing countries has a strong substantive element; it is not, as in the United States and Europe, a purely procedural requirement. This difference may have significant consequences (Houck 1999: 5).

Additionally, the law seeks to include the population in the protection of the environment through the development of interdisciplinary environmental education, public disclosure of actions and decisions, consultation with the public regarding these actions and decisions, and "adequate and sufficient access" to both administrative and judicial bodies to ensure compliance with the law and its provisions.

Article 7 provides for adequate funding of the various environmental protection programs: "The State will make provisions in its budget for the monetary appropriations necessary to meet to the requirements of programs related to the environment without prejudice of other agencies, bodies, and entities." The National Fund for the Environment is intended to fund projects aimed at the protection of and rational use of the environment and is to be administered cooperatively by the Ministry of Finance and Prices and Ministry of the Economy and Planning in consultation with Ministry of Science, Technology, and the Environment.

Because the strongest criticisms of Law 33 was for its inability to ensure compliance, the new law stipulates that any of the three following entities may demand the repair or indemnification of environmental injury: a) the office of Attorney General, b) the Ministry either acting on behalf of the public, or c) any person who has "personally suffered damage or injury." Reparation for environmental damage is focused more on the cessation of the damaging activity and repair and rehabilitation of the environment. Article 74 of Chapter 12 establishes the requirement of obligatory civil liability insurance to cover accidental environmental damage (Law No. 81 1997: Chapter 12). Thus the foundation for penalties, liability, and criminal responsibility is written into the law.

The new law on the environment deepens and strengthens the environmental impact assessment procedures that had been in place since 1995 extending them to include Cuban as well as foreign development projects. The types of domestic activities covered include road construction, agricultural, forestry, and aquaculture facilities, hospitals, airports, and industries.

According to Chapter 2 of Title 3 of the law, "The main objective of environmental planning is to insure sustainable development at different geographical levels, taking into consideration, in an integrated manner, environmental issues and their relationship with economic, demographic, and social factors with the aim of reaching the maximum harmony possible in the interrelations of society with nature." Land use planning is to consider the various characteristics of ecosystems, balancing human activities and settlements and environmental conditions, the "interdependence of humans and their environment," the environmental impact of new construction and the need for national defense (art. 22, Law No. 81 1997: Title 2, Chapter 2).

Much of the oversight of the various environmental functions of the ministry occurs in the Agency of the Environment (AMA) of the environmental ministry (CITMA). The current head of the agency, Dr. Jorge Mario García Fernández, oversees the implementation of environmental regulations, that gathering of environmental information, and environmental education. It administers three institutes or centers—the Institute of Research, the Center of Environmental Management, and the Center of Science and Culture—which in turn manage a number of smaller agencies/institutions. The Institute of Research manages the Oceanography Institute, the Institute for Ecology and Systematics, the Institute of Meteorology, the Institute of Geophysics and Astronomy, and the Institute of Tropical Geography. The Center of Environmental Management oversees the Center for Environmental Information, Management, and Education, the Center for Inspection and Environmental Control, and the Center for Protected Areas. The Scientific and Cultural Center manages the National Aquarium, the National Zoo, and the National Museum of Natural History. Various branch offices of the ministry report to the Science, Technology, and Environment Minister Doctor Rosa Elena Simeón.

Representing the environmental interests of the nation, the Environmental Agency establishes priorities for development projects along with representatives from international and national organizations in accordance with the responsibilities and requirements of the international treaties, pacts, and agreements to which Cuba is a party (AMA/CITMA). On the domestic front, the agency works in conjunction with the ministry's regional branches to coordinate the environmental management of the country, oversees the environmental inspection and licensing, and conducts scientific research and environmental monitoring through its various sub-agencies (AMA/CITMA). Two additional areas of particular importance for the agency include oversight of the national protected areas—the national parks and biosphere reserves—and the national environmental education strategy.

The Framework Law on the Environment prioritized issues and adopted positions, but much filling in of gaps was needed to translate these expressions of political will into actual rules and regulations. Some of this work was to take place in the articulation of a National Environmental Strategy. The National Environmental Strategy developed out of a process of negotiation between officials from the Ministry of Science, Technology, and the Environment and other ministries (CITMA/EAN 1999: 9–13). The document describing it begins by pointing with pride to the accomplishments the Cubans have made in their environment: "The balance of the environmental activity of the Revolution in these 37 years is outspokenly positive. The eradication of extreme poverty and its sequel in health and education terms; the improvements of the environmental conditions and of the quality of life [within] in an equity framework; the increase of the national forested areas, the statement of a set of protected areas and the proposal to integrate them into a national system . . . the use of science in the diagnosis and the solution of many environmental problems; gradual introduction of the environmental dimension in the national education system and the growing strengthening of the national environmental effort, are some of the achievements attained" (CITMA/EAN 1999: 9). The document goes on to list errors and deficiencies that still need correction, citing the need for greater public environmental awareness and knowledge through the expansion of environmental education; greater awareness and use of the latest scientific and technical advances; greater incorporation of the environmental dimension into policies, plans, and development programs; and finally, a legal system sufficiently integrated and coherent to implement the objectives of the framework law on the environment (CITMA/EAN 1999: 9–10).

CONCLUSION

One of the strongest criticisms directed at Cuba's old environmental law, Ley No. 33—from inside and outside of Cuba—was its lack of teeth to ensure compliance. But as CITMA officials, Orlando Rey Santos and Teresa

Borges Hernandez have argued in writing and personal interviews with me, the new law makes provision for penalties such as fines meted out on a case-by-case basis.[8] The law stipulates that any of the three following entities may demand the repair or indemnification of environmental injury: the office of Attorney General or CITMA either of which may act on behalf of the public interest as well as any person who has "personally suffered the damage or injury." Reparation for the environmental harm focuses more on the cessation of the damaging activity and repair and rehabilitation of the environment. "In discharging civil liability, preference will be given to actions directed to the rehabilitation of the environment." Article 74 of Chapter XII also sets forth the process for the development of obligatory civil liability insurance to cover accidental environmental damage.

The law also stipulates that certain acts prohibited by law under threat of imprisonment which "strike against the protection of the environment will be classified and penalized to the degree established by the law."

At least in terms of "teeth" there exists a foundation for penalties, liability and a criminal responsibility regime written into the framework law. Further inclusion of an environmental impact regimen sets the scene for environmental protection, analysis of projects that could negatively impact the environment, as well as the initiation of an educational process for business and government interests along with the legal standards and punitive regimen.

To guard against environmental degradation, the new law on the environment deepens and strengthens the environmental impact assessment procedures that had been in place since Law No. 77, the Foreign Investment Law adopted September 5, 1995. This law in essence affirms the ability of foreign investment to coincide with a sustainable development program for Cuba if it adheres to rational use of the natural resources and environmental protection. The law also states that potential foreign investors must participate in an environmental impact assessment, the procedures of which would be drafted by CITMA. Here is a rather surprising law that acknowledges not only the need for protection of the environment, but even more importantly the radical changes that Cuba will have to make to survive in a post-CMEA world. The new law appears to expand the environmental impact assessment process to include Cuban development as well as foreign investment-related development projects. The measure stipulates that all activity which could potentially produce significant environmental effects or "requires a specific control to conform with the requirements of existing environmental laws" would be required to obtain an environmental license from CITMA. The types of domestic activities covered in the law include road construction, agricultural, forestry and aquaculture facilities, hospitals, airports, and all industries such as mining, energy-generating, biotechnology, sugar, and similar consumable goods facilities. Again clearly, the Cubans are crafting laws that parallel the island nation's current transition from a centrally planned socialist economy to a mixed economy as seen in the rapid growth in the joint state- and foreign-owned industries, for example, energy and tourism. But it is necessary to continue to stress how very important the implementation process is in terms of the long-term success of this process in managing a real sustainable development policy for Cuba.

NOTES

1. González and García (1998) offer an excellent historical treatment of how natural resources, primarily, and also the environment was treated in Cuba's early history.
2. Columbus in his Navigation Diary cited in Del Risco 1999: 18–19.
3. EAN, p. 11. Laws are passed by the legislative body, the Cuban National Assembly. Decree-laws are proposed by state agencies, such as the Ministry of Science, Technology, and the Environment but must be approved by the Council of State, an executive committee that exists in permanent session. Finally the resolutions are adopted by individual agencies. Those resolutions are limited in jurisdiction to the adopting agency.
4. Acosta was also an active participant in the UNDP-UNESCO-UNED decontamination of *Bahia de Habana.*
5. Everson based this assertion on an interview with a law professor at the University of Havana in 1998.
6. As discussed earlier in this chapter, the environmental impact review process originated with the Foreign Investment Law. Houck and Speir noted in 2000 that the government seems to be committed to expanding and enhancing the environmental review process. My field research has shown the same through discussions with Cubans about this as well as witnessing the expansion of entities that carry out environmental impact assessments and reviews.
7. Houck 1999: 5. Houck participated in an environmental law exchange with the Cuban government from 1997–2000.
8. Also see Chapter XI, Articles 67–69, Law of the Environment, No. 81.

7

Opening Pandora's Box

We are not an oil producing country . . . the sea, the climate, the sun, the moon the palm trees . . . are the natural wealth of our country and we have to take advantage of them. —Fidel Castro 1959

The center of the country is also the patrimony of humanity at time when people throughout the world are eager to have fresh air, fresh water, to escape from the poison in the environment, the toxins that have accumulated. When people see these areas, they understand that resources are being created for humanity. Well-being is being created for humanity. At the same time, it will serve as a very important source of income for the country. —Fidel Castro 1989

The idea of traveling to Cuba invokes memories of escaping the cold grey of Northern winters to basking in warm tropical sun by day and gentle sea breezes in the sultry nights. Nights of *mojitos* and Cuban cigars, spent rolling dice on the casino tables or gyrating on the dance floor to the latest mambo beat at the Tropicana, complete the recollection.

Indeed over the ages, Cuba has had a love-hate relationship with tourism. There have been four major periods of tourism expansion in Cuba's history: 1920s, 1950s, 1980s, and the late 1980s.

The revolutionary government has at times turned to tourism, perhaps reluctantly, as a way to finance the type of economic and social development it wished to bring to the Cuban people. Indeed, in the very early years of the revolution, after Castro and his new government proposed closing down the famous gambling casinos arguing that they destroyed Cuban morals, hotel workers complained they would lose their jobs (Schwartz 1997: 196). Given that the industry employed thousands of Cubans in the hotels, restaurants, and clubs as musicians and other service employees, the revolutionary government had to weigh the question of unemployment, as well as the $2,000 per month and 20 percent profit from each casino, that could be used to help finance social development, with the ideological conflict over morals (Schwartz 1997: 197). Considering the protests of the hotel workers and the economic benefits, Castro changed his mind and decided the government would use the casino proceeds to develop a rehabilitation agency to create jobs to replace the gambling-related positions as well as prostitution (Schwartz 1997: 197). Despite these efforts, however, the tourism industry did not flourish in the politics of the Cold War.

The Cuban government in 1980 attempted to revive the tourism industry and served as a vacation spot for Soviets in particular as well as some Europeans. There were modest gains (about two hundred thousand tourists in 1982), but it was not until the collapse of the socialist bloc in 1989–1990, that tourism efforts began in earnest (Schwartz 1997: 205). Cuba needed, as it had in its earlier efforts, to develop tourism in conjunction with foreign investors. At that point, the Europeans were especially important. Moreover the tourism industry in Cuba needed some updating to be competitive in the Caribbean hotel and restaurant business and again the Europeans would be helpful (Schwartz 1997: 206).

A major milestone in the development of the Cuban travel industry came in 1991 with the island's hosting of the Pan-American Games. Schwartz argues that the government hoped Americans would see Cuba from their living rooms and would soon press their government to allow them to see the island in person. That year also saw the release of a promotional campaign in the form of the August 1991 National Geographic article on Cuba that would coincide with the start of the games (Schwartz 1997: 209). This time though the government would not take a chance on using gambling to boost its tourism industry.

The naissance of foreign investment in the tourism industry began that year as well with the construction of the Melía Varadero, a 390-room seaside hotel with the Spanish government on what is one of the country's premier beaches.[1]

FOREIGN INVESTMENT

That year was also when the foreign investment law, Law No. 77, was passed. The 1995 foreign investment law was approved to consolidate the foreign investment process based on the lessons learned by the nation in recent times, according to a small group of well-known Cuban economists working within the country (Carranza, Gutiérrez, and Monreal 1996: 51). The economists stress the need for periodic review of the overall policy as well as clear delineation of the specifics of which sectors can receive foreign investment and the process for foreign investment via the oversight of and control by the country's highest legislative body, the National Assembly of Popular Power (ANPP).

Foreign investors could participate in Cuba's economy in one of three ways: first, in joint ventures between national investors and foreign investors, creating a Cuban commercial company; second, through international economic association contracts where national and foreign investors participate in a joint venture "even without the establishment of a legal entity distinct from each of the parties, and third, as a totally foreign company, defined as a commercial entity with foreign capital, without the involvement of any national investor" (Art. 2, Law No. 77).

The new law expanded the financial collaboration Cuba engaged in with other, mostly Western European governments as well as the Canadian. Canadian companies are quite involved in business ventures with Cuba. Canadian businessman Ian Delaney, who has befriended Castro, speaks highly of the bi-national business arrangement, "The Cubans are terrific people to do business with. . . . For one thing they're totally incorruptible. We do business all over the world, but there's no place we like better to negotiate deals than Havana. They're trying to develop a system that has few parallels. Perhaps Canada, which is really a social democracy, comes closest" (Delaney quoted in Newman 1995: 45).

The joint ventures expanded after 1992 and nearly doubled between 1994 and 1995, according to Azicri.

In early 1995, 290 projects were under consideration, and by the summer 212 new partnerships were in full operation, increasing to 260 a year later. They were limited initially to ten countries, but soon expanded to include fifty-three nations. Investors from Spain, Mexico, Canada, England, France,

Table 7.1. Foreign Investment Projects in Cuba by Country, 1990–1995

Country	*1990*	*1991*	*1992*	*1993*	*1994*	*1995*	*Total*
Spain		3	9	10	14	10	47
Mexico		2	3	3	4	1	13
Canada			2	8	16		26
Italy			1	5	4	7	17
France	1		3	5	2	2	13
Holland			1	2	3	3	9
Rest of Latin America		2	3	11	9	4	29
Rest of world	1	4	11	16	22	4	58
Total	2	11	33	60	74	31	212

Sources: Azicri 2000: 154; and Comisíon Económica para la América Latin y el Caribe-CEPAL, *Cuba: Evolucíon Económica durante* 1995 (NY: United Nations, 1996), 22, based on figures provided by the Cuban Ministry of Foreign Investment and Cooperation, May 1995.

Table 7.2. Foreign Investment by Country, 1995–2001 (Miles de pesos)

Country	*1995*	*1996*	*1997*	*1998*	*1999*	*2000*	*2001*
Spain	446,293	570,393	667,372	749,279	881,180	893,338	882,503
Mexico	400,337	361,267	392,779	387,930	346,821	337,815	315,097
Canada	451,071	455,464	514,402	551,215	568,827	589,019	590,217
Venezuela	237,671	334,221	437,796	387,728	464,765	912,409	937,197
Italy	165,087	178,855	272,796	281,182	303,346	330,800	307,343
France	263,391	252,884	304,691	361,491	326,854	332,287	330,771
Holland	164,584	183,723	133,114	132,205	164,576	205,502	400,783
Russia	251,302	616,086	415,265	490,135	427,321	435,877	484,185
China	359,697	275,857	345,490	418,351	481,732	524,301	618,564
Japan	107,007	109,739	154,539	95,952	125,631	111,102	109,567

Source: data drawn from *Anuario Estadístico de Cuba* 2001.

and Holland led foreign investment in Cuba. The 1996 U.S. Cuban Liberty and Solidarity Act, more commonly known by the names of its congressional sponsors as the Helms-Burton Act, sought to stop this essential component of Cuba's economic recovery (Azicri 2000: 153).

Ironically, Cuban officials said the Helms-Burton law did not stymie these types of investments in Cuba. "Since the . . . Helms-Burton law, there has been more foreign investment in Cuba than in all of last year . . . investments have not ceased" (Azicri 2000: 154). Estimates for 1998 put the joint venture figure at between 330 and 360.

FOREIGN INVESTMENT IN TOURISM

Tourism, along with mining, now dominates the Cuban economy. Azicri states that tourism, as of 1995, was the island's primary source of hard currency even surpassing remittances from relatives living abroad (Azicri 2000: 157). In 2001, nearly 1.8 million visitors came to Cuba, including a surprising number of Americans, nearly 79,000 (Oficina Nacional de Estadísticas 2002: 252–253). The Cubans surpassed the two million visitor mark in late 2004 (Rojas Aguilera 2004). The fact that this type of economic expansion must be accomplished while considering the protection and rational use of resources stands out.

While the law allows for foreign investors to own up to 100 percent of a business in specific economic areas, according to Azicri, most of the most notable investments have been made as joint ventures. Castro though has indicated state-majority ventures would be the rule and completely foreign-owned endeavors would be the exception (Castro in Navarro 1995).

Specifically the Europeans have moved in this investment area, but the Canadians and other Latin American nations have as well. The Italian company Siata is one such example. Siata is working with Cubanacán on four hotels, two of which are in Varadero and two are in Cayo Coco (Azicri 2000: 155). Siata's president said, "Cuba's future lies in tourism because it has all the right characteristics: first of all, its people, and second, the confidence which this country inspires" (Azicri 2000: 155).

Spain's Sol Melía company dominates the hotel industry in terms of partnership with the Cuban government. Sol Melía runs twenty-two facilities with about forty thousand rooms in coastal areas such as Varadero, Cayo Guillermo and Cayo Coco in the north, and Cayo Largo in the south as well as in Havana and other cities (Granma Mar. 18, 2004). To compare there were only about 189 total hotel facilities and some thirty thousand rooms in the country in 1999. The other primary international companies involved in joint venture agreements with Cuba are Super Clubs (Jamaica), Accor (France), and Barcelo (Spain). The Cuban companies that work with these foreign partners are Gran Caribe, Cubanacan, Habaguanez, and Gaviota (Jebodsingh 2001).

The Canadians have also invested in the Cuban tourism industry. In 1998, Sherritt of Canada expanded from its nickel and oil exploration interests to tourism, real estate, agriculture, and sugar as well as finance and communications (Azicri 2000: 154, CUBANEWS Dec. 1995: 7). In addition to Sherritt's investments, Varadero's airport was built by a Quebec company and the new terminal at Havana's Jose Martí airport was built by an Ontario firm (Kirk and McKenna 1997: 159). In fact, the Canadians note their relationship to Cuba's tourism ranges from the food and beer served, to the music equipment used in entertaining guests; all hail from Canada. One Canadian tourist consultant quipped, "Doing business in Cuba is a license to print money for Canadians" (Kirk and McKenna 1997: 160).

In 2001, the Cuban business weekly Opciones estimated that the tourist industry accounted for 40 percent of Cuba's hard-currency earnings that year, up from 4 percent in 1990. Cuban Tourism Minister Manuel Marrero Cruz reported tourism grew by 8 percent in 2003 (Rojas Aguilera 2004) with an increase of 13.2 in 2005 (Havana Journal Jan.30, 2006).

ENVIRONMENTAL IMPACT OF TOURISM

Passage of the new foreign investment law necessitated formal processes for environmental impact assessment to ensure that the quantity and character of the economic development did not injure the nation's natural patrimony.

The environmental law, in Chapter 16, addressed how foreign corporate participation would impact the country in terms of the environment. The law stipulated that foreign investment is "conceived and stimulated in the context of the country's sustainable development, which implies that during the course of the investment, environmental conservation and the rational use of natural resources are carefully undertaken" (art. 54). In this law, we see a clear process as to how a proposed foreign investment project would be analyzed in terms of how it could impact the environment and natural resource base. The Ministry of Foreign Investment and Economic Cooperation submits the proposals to the Ministry of Science, Technology, and the Environment (CITMA), so that the agency can determine if an environmental impact assessment needs to be undertaken. Also at this point the ministry would grant an environmental license for the project along with the control and inspection program required under the Framework Law of the Environment.

Responsibility for adherence to environmental legal standards does not end there. If a representative or a company injures the environment, then according to Article 56, s/he is "obliged to reestablish the previous environmental situation, repair the material damage and indemnify the injured parties." To deal with the impact of tourism development on the environment, Cuba made an exceedingly important step by adopting an environmental review process.

In addition to the foreign investment law itself, there are encouraging signs the Cuban have implementing so as to more rationally manage the environment as it develops its tourism industry. First, the Cubans have invested heavily in these areas and indeed the majority of the money has gone to the coastal areas (Lindeman et al. 2003: 597). Second, these areas are subjected to the environmental impact analysis (EIA) process. These EIAs help inform officials on whether the proposed projects should be granted environmental licenses required for development. About one-third of the projects proposed to the Cuban inspection and environmental control agency require environmental impact analyses (Lindeman et al. 2003: 600). While the total number of permit applications, both domestic and foreign initiated, to the inspection agency has tripled from 1997 to 2000, there is evidence that not all projects are granted environmental permits (Lindeman et al. 2003: 602). For example, "in 2000, 736 out of 748 license requests were granted" and many of the proposed projects had to undergo significant alternations to avoid or mitigate potential environmental damage before being granted that license (Lindeman et al. 2003: 602). According to one U.S. legal and scientific analysis sponsored by Tulane University Law School, "this general ratio is not substantially different" for similar coastal development permits requested in the United States (Lindeman et al. 2003: 602).

Under Cuban law, the environmental ministry can require such projects be altered so as to minimize environmental impact and even can deny permission for such projects to move forward as I discussed in the previous chapter. My research found a number of Cuban environmental impact analysis consulting businesses have developed in response to the environmental law.

There is also evidence that the environment ministry does use its teeth to enforce the environmental law. For example, in 2001, Cuba sanctioned more than 500 individuals and firms for law violations, fined 178, issued 308 compliance order, and temporarily or permanently shut down a number of businesses which failed to comply with the environmental regulations (Peláez 2002: 3).

In addition to the environmental impact analysis regimen, Cuba has also crafted zoning laws regarding seaside tourism facilities aimed at protecting the fragile coastal zone. Cuba's coastal zone management law, Decree-Law No. 212 passed in 2000. This law stipulates that development be set back out of the inland boundary in the following manner: sand, rock and gravel beaches, coastal cliffs and river mouths 20 meters, and beaches with sandbar or dunes, low coasts with mangroves, and other fragile coastal zones 40 meters (Section 1, Article 4, Decree-Law 212, 2000).[2]

Under current Cuban law, siting of a project happens before environmental impact analysis, but this is not unlike what frequently happens in the United States. Whittle et al. recommend a thorough environmental impact analysis be conducted early in the process so as to reveal potential harm as well as location of alternative sites before the project's exact location be finalized (Whittle et al. 2003: 575).

ECOTOURISM

A number of countries, like Costa Rica, have adopted ecotourism principles in an effort to build tourism while preserving the environment. "Ecotourism is a form of low-impact, low-technology, and low-capital tourism development, explained Megan Epler Wood, president of the Ecotourism Society, based in Vermont (Simpson 1999).

Cuba contains some sites where ecotourism is developing or could be developed. One example of ecotourism is located in the Sierra del Rosario Biosphere Reserve in the community of Las Terrazas and featuring the environmentally principled hotel La Moka. The surrounding land was originally a coffee growing area founded in the late 1700s by French who left Haiti. Because the French owned the desirable land, the locals had to subsist on the less fertile and more vulnerable slopes. In the early 1960s, the government led a 5,000-hectare reforestation of the area and constructed a model village for the residents. There is a museum in the area depicting the reforestation efforts as well as an ecological center. This is a relatively small-scale project that has a much lighter footprint than the large-scale hotels Cuba has focused on constructing in other areas.

Another area with only limited development is the Zapata wetland. The one million acre Ciénaga de Zapata, the ecological, albeit healthier, twin of Florida's Everglades, is considered the most ecologically important wetland in the Caribbean. The wetlands are the winter home for a multitude of migratory birds as well as twenty-two endemic birds and its famous Cuban crocodile. Zapata is an important bird watching location yet it is also a fragile ecosystem. Access to the wetlands is monitored and limited at this time. One site that is used for environmental education is the breeding facility, *Criadero de cocodrilos*, for Cuban crocodiles. Visitors can view crocodiles as they grow from hatchlings to juveniles before release as well as learn about the importance of this top predator for the health of the wetlands. The environmental ministry also has an interesting ecological site in the area.

In Viñales Valley in western Cuba, a number of potential sites could be used for ecotourism with the addition of improvements to its environmental educational aspect. *Valle de Viñales* is a spectacular valley surrounded by limestone buttes, called mogotes. Guides offer some cursory geological information, but this is an area which could be enhanced, as well as discussion of the various birds and animals that inhabit the area.

An interesting site in Viñales is the Cuevas del Indio, a four-kilometer long cave where the remains of Cuba's original inhabitants were found. Visitors walk through a series of rooms that sometimes reach 135 meters in height, and even board a boat for a journey down the green river. The guides provide information on some of the bats and blindfish that occupy the cave, but much more could be added as to both cave life as well as the intricacies of cave environments. Also more information on the archaeological and anthropological aspects of the Cuevas del Indio would help round out an ecotour.

Cuba's coastal areas, ranging from Varadero to Cayo Coco and Járdines del la Reina could also be a site for ecotourism if developed properly. These areas are home to wildlife like flamingoes and iguanas as well as the multicolored life around the reefs. Again caution is needed as these areas are developed. But the Cubans are quick to point out that they have not really begun developing ecotourism. Environmental Ministry officials told me in 2003 that Cuba had not yet developed an ecotourism industry. Cubans are troubled over the concept of ecotourism. Miguel Vales García, director of the national center for biodiversity, has expressed such concerns. "There is a general idea that ecotourism is putting a hotel inside a beautiful natural area. This is not ecotourism," he says, adding, "Ecotourism is increasing the knowledge of people without destroying nature" (Vales García quoted in Simpson 1999). Vales worries about the environmental impact of even ecotourists to a fragile site, speculating that if a hotel is built in a fragile site, the impact could have devastating effects on the plants and animals that inhabit the area (Simpson 1999).

CONCLUSION

Cuba has the potential to build an effective ecotourism sector to its burgeoning tourism industry. The principles of ecotourism, like tourism itself, must be founded on social justice principles and environmental protection. The social justice concerns can be addressed by paying the workers in the industry a living wage. Indeed in Cuba, tourism workers often earn more than ordinary Cubans. These situations of inequity should be rectified. Also the development of the tourism industry must treat the employees in a respectful and culturally appropriate manner. In terms of the environmental protection, Cuba has already drafted a strong framework environmental law as well as the environmental impact analysis process. The framework law and the EIA process should be updated and strengthened as needed. Cuba must continue the balance of the development of the tourism industry with the larger concerns of social justice and environmental protection.

It is encouraging that the Cuban environmental ministry has sanctioned, fined, and even halted construction of projects because they did not comply with the environmental regulations. It is also a hopeful sign that Cuba has created a new coastal management law aimed at protecting the fragile coastal areas where much of the new tourism development has occurred. This pattern of enforcement and updating of laws is important for the island as it continued to balance economic development, social justice, and environmental protection.

One aspect that Cuba is developing which will be helpful in the development of ecotourism and tourism in general is the environmental education sector. Public understanding of the necessity for protection of the natural world and its connection to Cuba's short-term and long-term viability and social justice commitments will make compliance with the environmental regulations more feasible. We will examine the development of environmental education and its impact on Cuba's road to sustainable development in the next chapter.

NOTES

1. The hotel was the first to receive the Environmental Acknowledgement award granted by the Ministry of Science, Technology, and the Environment nearly a decade later (Borges 2003).

2. Law professors from the University of Tulane Law School, as well as scientists from the Center for Marine Conservation, held working group sessions with CITMA officials in May 1999 to work on a preliminary version of the decree-law.

8

Greening the Curriculum

If we want a child who takes care of the garden, teach him to sow a garden . . . and no one will have to spank him with a stick, teach him to plant a tree and no one will have to punish him for destroying trees. They destroy what they do not understand. Those who destroy do not understand what it means to create.
—Fidel Castro Ruz

A key challenge to the ability of a society to enforce well-intentioned and even far-reaching environmental laws such as those that were passed in the late 1990s in Cuba is making people aware of their role in the enforcement process. The Cuban case demonstrates the importance of having an informed and educated population in understanding and thus adhering to a given body of environmental laws. Further, if the citizenry is expected to participate in decision making about the management of the nation's resources, it must have the necessary educational background to make informed decisions.

Cuba devised its national environmental education strategy the same year as Law 81 and the National Strategy of Environmental Education were developed within the framework of the National Environmental Strategy (CITMA/ENEA 1997: 12, EAN 1999: 6). Its goal was the education of students, professionals, and workers, and the general public. The strategy sought to introduce an environmental dimension school system at both primary and secondary levels and higher education (EAN 1999: 6). Environmental education of leaders, professionals, and workers was to be carried out via programs undertaken at the sectoral level. Finally, public environmental education was to be realized through public education campaigns and mass media coverage of specific issues as well as strengthening institutional capacity (EAN 1999: 6).

The Cubans take as their inspiration for this program a passage from the Rio Summit's Agenda 21 on education and the environment:

> Education is critically important to the promotion of sustainable development and the ability of the populations to tackle questions about the environment and development. To be effective, education about the environment and development must be concerned with the dynamics of the physical and biological world, socioeconomic realities, and human development (which may include the spiritual development of humans), in an interdisciplinary approach utilizing academic and non-academic methods and effective means of communications.[1]

Cubans see environmental education as a powerful instrument to achieve changed attitudes toward the environment and halt the alarming deterioration of the natural world.

The environmental education projects include empowerment and consciousness-raising about the protection of the environment. The strategy provides for collaboration between local Poder Popular councils in the facilitation of "community training and environmental self-management tailored to the necessities and conditions of each locality" in terms of problem solving and decision making (art. 54, Law No. 87 1999: 20). It situates the educational process within Cuban realities by stressing the strengthening of Cuban cultural identity, through care for the nation's patrimony, its resources, and its sovereignty (ENEA 1997: 7).

In addition to facilitating local-based participatory problem solving, the strategy is designed to infuse Cuba's youth with environmental consciousness by introducing environmental issues and concerns in school curricula and expanding educational preparation in careers directly related to environmental protection, rehabilitation, restoration, and the monitoring of environmental programs.

Law 81 outlines the collaborative approach between the Ministry of Science, Technology, and the Environment, the Ministry of Education, the Ministry of Higher Education, the Ministry of Culture, and other state agencies required to craft a national strategy for environmental education (art. 46, Law No. 81 1997). While there is clearly a focus on "greening the curriculum," environmental education is not to be limited to the classroom nor to the hours of instruction. Cubans working in recreational, scientific, and cultural institutions are called upon to develop a variety of activities to raise levels of environmental awareness in the general public (art. 51, Law No. 81 1997: Chapter 7). In addition, the mass media are encouraged to include environmental educational themes in programming to draw out "themes that will foster greater information and understanding in the population about the complex interrelationships and links between the processes of economic and social development and environmental protection, fostering a higher environmental culture among the public" (art. 53, Law No. 81 1997: Chapter 7).

The law also includes provisions for state agencies and bodies to participate in the drive for environmental awareness with special focus on protection of and rational use of resources during productive activities as well as in worker training. Here we see the diffusion of the concepts of environmental management and rational use of resources throughout the productive sectors of society to further cement this new approach to development. The National Environmental Education Strategy was the necessary step in making the normative vision expressed in the framework Environmental Law a reality.

Cuban environmental education specialists trace the environmental educational process to the global move toward increasing awareness of environmental issues by various agencies of the United Nations, particularly through UNESCO's Man and Biosphere Program, and international educational organizations in the 1970s (Roque 2002).[2] These efforts were characterized by a focus on education as the catalyst for change in the way humans interact with their environment. Educators around the world began critiquing the models of development that were being embraced by the industrialized world and diffused to the Global South, and out of these discussions emerged "the historic conclusion that the environment and development are not incompatible" (Roque 2002: Part 1). The International Education Program, oriented to the protection of nature, was formed in the 1970s, and this program also helped shape the development of Cuba's environmental education programs (Roque 2002: Part 1). The 1992 Rio Summit was a watershed moment. Often called the Earth Summit, the June 1992 meetings drew representatives from 179 nations as well as seven thousand NGOs and thousands of private citizens. The participants put forth principles on global environmental governance in what is called Agenda 21, setting forth a plan of action to implement these goals in the twenty-first century. At the Rio conference, the majority of nations (the United States not among them) agreed that underdevelopment and poverty were the main obstacles in reaching higher levels of sustainable development. Rio was a turning point in the minds of many environmental educators and surely those in Cuba. The plan of action of the conference called for the facilitation of "access to education on the environment and development" within the context of social education for all sectors of society, school children to adults, to reorient the populace toward sustainable development (Roque 2002: Part 2).

In their experiences at conferences such as the Ibero-American Congresses and the World Conference on Education and Society, Cubans came to see that environmental education was generally viewed as divorced from the historical social, political, and economic realities of Latin America and the Global

South and that a new direction, incorporating a social component into the broader concept of sustainable development, should be pursued (Roque 2002: Part 2). She concludes environmental educators help transform how people think and interact with one another and with nature through an educational process centered on values regarding the environmental aspect of development and help reorient us to the aims of humanity.

Cubans view education as a culturally and economically grounded process with the goal of the full development of the "most transcendental qualities of the person . . . in terms of knowledge, capacities, feelings, convictions, will, and values" (Roque 2001). Since the 1970s, there has been an increase in popular participation in education, and high-quality education is an important value as well-documented by Lutjens.[3] Education is viewed as a path to the creation of a new socialist person—"an integrated, harmoniously developed personality, so that each individual becomes a conscious and active builder as a beneficiary of the new society that is being built" (Ministry of Education 1984: 123 cited in Lutjens 1996: 82).

The Cubans point to the national hero José Martí as a pioneer in the protection of the environment, and a source of "an ethics rooted in the respect and protection of nature" that constitutes a basic principle of the educational system (ENEA 1994: 4). The pioneer Cuban environmentalist Antonio Núñez Jiménez has described this tradition as a "culture of nature" (Núñez Jiménez 1998).

The National Environmental Education Strategy analyzes the linkages between the expansion of global capital and environmental degradation. Globalization is viewed as having a model of development that justifies the "deterioration of ecosystems, the loss of the biodiversity, and the unjust distribution of the wealth, and therefore increased poverty is intrinsically tied to processes of cultural homogenization, oriented to the export of the untenable patterns of consumption that characterize the economically developed societies and components of the environmental problematic" (CITMA/ENEA 1997: 3). The strategy views education as an agent of change for the reorientation of humans to a harmonic relationship with nature in order to achieve the sustainable level of development that will result in a better and more equitable life for everyone on earth. The education strategy recognizes that the conditions under which Cuba has existed because of the blockade have contributed to the development of a creative and collective pursuit of alternatives based on science and a practical approach to problems (CITMA/ENEA 1997: 3). Identifying a lack of sufficient environmental consciousness on the part of Cubans, the strategy is aimed at incorporating the environmental dimension into everyday life.

This strategy shows how Cuba, in the face of adversity, sought self-sufficiency and the careful use of resources out of necessity. It also clearly demonstrates the evolution in thought toward the island's natural resources and how the respect for nature coincides with the country's commitments to social justice and is much more aligned in purpose than first realized in the early years of the revolution.

According to the document, "the rational management of natural resources and practices of equity and social justice are historical responsibilities that we have because of the linkage between humans and the environment to which we belong" (CITMA/ENEA 1997: 8). Sustainable development is "conceived of as a process of creating the material conditions, cultural and spiritual, that provide for an increase in the quality of life for society, with equity and social justice, based on a harmonious relationship with social and natural processes focused on both the current generation and future generations" (CITMA/ENEA 1997: 10). The object of sustainable development is the human being, and its goal is the satisfaction of humanity's real needs as well as the equity of all.

Basic principles of the national environmental education strategy include the following: (1) recognition of the political and ideological character of environmental problems, the link between the global expansion of capital and hegemonic interest of the industrialized world, and unsustainable levels of consumption, resource use, inequalities of wealth, environmental degradation, and social injustice; (2) environmental ethics, an ethical perspective centered around the recognition of the interrelatedness of life on the planet and the right of all species to be preserved; (3) a systematic and interdisciplinary approach to environmental problems involving the cultural historical patrimony, social relations, socioeconomics, and biological systems; (4) recognition of the dialectical relationship between the global and local levels of environmental problems and the interdependence of environmental problems and development; (5) attention to the historical nature of environmental problems; (6) the pursuit of international collaboration while preserving the independence and sovereignty of the nation; (7) recognition of the link between instruction, training, and education and a dialectical vision of knowledge, capacity, and values; (8) advocacy of a participatory educational process; (9) respect for both scientific and traditional knowledge; and (10) recognition of the dialectical unity of theory and practice.

To accomplish these lofty goals, Cuba is working with agencies within the country and international governing and nongovernment organizations outside the island, such as Save the Children and UNESCO.

One project, sponsored by the Center for Environmental Information, Management and Education (CIGEA) and Save the Children and aimed at young people from eleven to twenty-one, is designed to encourage participation in environmental decision making as citizens or through work as technicians and workers (CITMA/CIGEA). Another program, co-sponsored with UNESCO, focuses on raising awareness about the Caribbean Sea and humans' relationship to it (CITMA/CIGEA). The program educates students about the natural wealth and principal environmental problems of the Caribbean and helps develop the skills to resolve them. Another program focuses on land and marine resource conservation on the archipelago of Camagüey (CITMA/CIGEA). The students learn about the endemic species of the various ecosystems as well as the many archaeological and paleontological sites in the area.

Environmental awareness is fostered in children and young people through festivals, tours of environmentally sensitive areas, tree plantings, public announcements, videos, and writing and art contests (CITMA/CIGEA 2000). Adults outside the formal system of education are targeted by a television series called *University for All,* in which environmental education takes place alongside instruction on history, languages such as English and French, and the arts.

This program covers humanity's relationship between the environment, soil degradation, air pollution, biodiversity loss, and deforestation and includes a clear and accessible discussion of the nation's environmental law and its various programs for conservation and a discussion on cleaner production processes. The theme of the final episode is Antonio Núñez Jiménez's phrase "Toward a Culture of Nature," and it highlights the need to develop a responsible way of interacting with the natural world. It examines the various participants in the process—the family, the mass organizations, and the media—and shows how the adoption of "an environmental culture" can enhance the quality of life for all. The program uses video clips and graphics to complement the narration and the information is presented respectfully and straight-forwardly.[4] The environmental nongovernmental organization founded by Núñez Jiménez, Foundation for Man and Nature, worked with the government on some of the episodes.

JORGE RAMÓN CUEVAS

From 1988 until his untimely death in August 2000, the scientist Jorge Ramón Cuevas played a formative role in the minds of Cubans when it came to environment.[5] His memory remains constant in the hearts of many Cubans who are often moved to tears by the mention of his name. Cuevas hosted a television show called *Entorno,* which looked at nature around the world. He was a soft-spoken "televangelist" for the preservation of Cuba's natural world. He designed his show so that Cubans were exposed not only to the diversity of the island's natural resources, but also the many creatures and plants of the world. Cuevas told me he believed that if Cubans saw the beauty in what they had they would value it and take care of it. Environmental education and the development of a Leopoldian-style land ethic is what Cuevas thought was the necessary motivation for the preservation of

Cuba's resources.[6] He focused on opening people's minds and beginning a dialogue. He was determined to reach Cubans at every level of awareness and education. Although he did not mention the Brazilian educator Paolo Freire by name, he obviously employed Freire's pedagogy of reflection and action both in his television series and in his life.[7] When I first spoke with him in 1994, Cuevas had recently aired a segment that dealt with the use of earthworms in the composting of kitchen scraps. Many viewers had responded, some positively and some negatively. What was important to Cuevas was not that viewers reacted in one way or another, but that they considered the material presented. "I liked it when many people spoke," he said. "It does not matter whether they like it or not; what it is important is that they think and talk about it." Here he was illustrating the participatory way in which citizens are being educated—a process about the environment that must occur before full participation in policy decision making regarding the natural world can begin. "If you give people a solution, then it goes right through them. But if you leave the questions open, then they will think and talk about it and find a solution," explained Cuevas. Without environmental education, decision making will be vulnerable to distortion by the politics or economics of the day as is so often the case in other societies. Furthermore, participation in environmental decision making equips the public to understand, accept and abide by the decisions in question.

HAVANA METROPOLITAN PARK

The Havana Metropolitan Park rehabilitation project links participatory environmental education with sustainable development and environmental protection in a striking way. The 700-hectare Almandares River Watershed of which the park is a part was largely developed with housing, industry, and agriculture. The river that runs through much of the park had become quite polluted with industrial waste, agricultural runoff, and sewage from local communities, and the pollution flowed into the already polluted Havana Bay. In 1989, a working group was formed to revitalize the park, and in 1994, the project was expanded in conjunction with the Canadian International Development Agency and the Canadian Urban Institute to include restoration of the river and renovation and sustainable development of the surrounding watershed area (IDRC). Unless actions of the area's residents and industrial and agricultural neighbors were altered, restoration efforts would be effective only until the next spill of chemicals or sewage runoff. The project was aimed at reversing deforestation, increasing the practice of urban organic agriculture, cleaning up the area and the river, finding alternatives to local sewage and solid waste problems, renovation of housing, the development of recreation opportunities, environmental education, and the encouragement of self-sufficiency and participatory models of development.

Efforts were made to involve the communities in the restoration effort through interaction with the nine popular councils that represented the forty thousand local residents living in the watershed area. Members of the Metropolitan Park Association and employees worked with community leaders to develop the environmental plan, and asked children at the primary and secondary levels "help move the hearts" of their parents, and other residents. They assisted communities in developing programs of self-sufficiency in waste management, recycling, and overall development, and as communities reached a particular level of development, "weaned" them from the daily assistance and moved to the next community.

Park employees also served as consultants to industries and communities in the area of water pollution. Yociel Marrero Baez, project manager for the Almandares River cleanup, described the process of consciousness-raising that occurred in the meetings with community members.[8] When asked about the source of the river's pollution, he said, many pointed to industrial waste, but when household waste was mentioned, there was a realization of their own role and the control they had over their impact on the health of the river.

Another aspect of the project that had been involved the construction of a wetland in what was formerly a shantytown hastily constructed without official planning or permission. The settlement lacked, among other things, proper sewage connections, and so waste from the homes flowed downhill into the river. The construction of the wetlands was designed to rectify this situation from an environmental standpoint, and improve the lives of the area's inhabitants by cleaning the water sufficiently to permit its use for irrigation of individual and community gardens. The sludge generated is to be used as a fertilizer for the gardens.

An analysis of the entire project pointed to the importance of conveying a more holistic view of the area's problems and stressed the importance of continued community participation in the process. It also pointed to the significance of the Cuban experience in facilitating the participatory process and contributing to the long-term success of such projects.

Further the example of the metropolitan park shows us the importance of government and community cooperation in crafting environmentally sustainable local-based models of development. We see examples of the results of a society's commitment to social and human development as well as environmental education as necessary prerequisites for the transition to a more sustainable society.

By participating directly in environmental decision making, citizens can begin the transformative process to create a sustainable society. These changes include consideration of societal and individual needs as well as those of current and future generations; the way in which people live and their self-actualization; how agricultural and industrial development is conducted; the balance of living standards against true needs and false needs, and being empowered to full participation in the attainment of a truly sustainable society. A state cannot simply allow people to make uninformed decisions about environmental protection and resource use as power differences and lack of knowledge hamper the abilities to make decisions to benefit all in the society, as well as to protect the natural resource base.

On the other hand, a state cannot dictate laws, policies, and regulations on environmental protection without public participation in the decision-making process as the population will fail to recognize their impact on the environment resulting in antagonism and disrespect toward the law. Cuba appears to be taking a third road. adopting and implementing far-reaching environmental protection laws and has made a strong commitment to environmental education in both the formal and informal sectors of society.

These efforts lay the groundwork for future full and meaningful participation in environmental decision making. Just as Cuba vowed and succeeded in eliminating illiteracy so people could fully participate in society, it is fostering awareness about the scientific, environmental, economic, and social implications of the human impact on the environment. The challenge for Cuba is to translate the results from the environmental education project into meaningful full public participatory efforts in the governance, protection, and utilization of their natural resource base to fully realize the dream of a sustainable society.

Empowerment, individually and collectively, has been an important ingredient in the Cuban revolutionary prescription since its inception. Similarly, we see the necessity for empowerment in terms of participation in the crafting of a sustainable society. This is illustrated quite clearly in the metropolitan park example. In addition to the knowledge base required for decision making, empowerment must also come in the participation of the design, construction, and maintenance of the community in which one resides. Given the history of environmental difficulties over NIMBY (not in my backyard) concerns in other cases, participation in the development of one's own community is fundamental to the empowerment endeavor in the construction of a sustainable society.

NOTES

1. Chapter 36, "*Fomento de la Educación, la Capacitación y la Toma de Conciencia,*" de la Agenda 21 cited in ENEA: 1.

2. Roque is an environmental education specialist who works for CIGEA/CITMA. Roque offers an excellent historical treatment of this topic.

3. See Sheryl Lutjens, *The State, Bureaucracy and the Cuban Schools: Power and Participation,* Latin American Perspectives Series No. 17, Boulder, Colorado: Westview Press, 1996, p. 82.

4. Author's observations, June 2003.

5. Cuevas was killed in an automobile accident while in the field producing one of the episodes. In 1994, he estimated he had created more than four hundred programs, which aired weekly on television.

6. Aldo Leopold advocated the development of a land ethic in which humans viewed their relationship to the natural world as that of steward. For more, see Leopold's *A Sand County Almanac.*

7. Freire is known for his pedagogical methodology of reflection and action. For more, see *Education for Critical Consciousness,* Paolo Freire, 1974.

8. Field notes, June 2003.

9

People, Nature, and Development from the Cuban Point of View

This infallible theory of sustainable development and the economic model consistent with it will remain "missing links" unless the starting point is taken to be the human being as center, object, and subject of development. —Gina Rey

It is apparent from the above statement that one country in the midst of economic and political turmoil has made significant progress toward sustainable development in both social justice and ecological terms.

The foundational principles for sustainable development as set forth in the World Commission on Environment and Development report (known as the Brundtland Report) include meeting the basic needs of people, particularly the poor; recognizing technological, social, and natural limits to providing for the needs of current and future generations; developing the intellectual, problem-solving, and creative energies of humanity; and ensuring equity between current and future generations and, to a lesser extent, between members of current generations. For Cubans too, sustainable development is found at the intersection of social justice and environmental protection. As the Brundtland Report argues, people cannot simply be viewed in terms of numbers or only as consumers. They must have their basic needs met before their intellectual, problem-solving, and creative development can commence. Cuba is following a path of sustainable development by meeting the basic needs of all its citizens in food security, education, and health care; facilitating the development of the whole person by its commitments to education, as well as athletic and artistic expression; creating a wide-ranging, comprehensive environmental protection legal framework; developing an environmental consciousness through society-wide environmental education programs and efforts to develop an environmental ethic; working toward self-sufficiency in food, energy, and medicine; and enhancing public participation in local-based development and environmental decision making and management.

The transition to the pursuit of sustainable development was not entirely deliberate, but it was forged within a socialist framework out of what some might call the need to survive and others might consider the result of a dialectical process. What the Cuban case offers to students of environmental protection, sustainable development, and social justice in the Global South is the challenge of creating a society that places humans both current and future generations at the center. Cubans are creating "rational" modes of living that do not impinge upon their country's fragile ecological balance.

When Western social scientists use the term "rational" the context is macroeconomic theory's notion of economic rationality and rational choice. They employ rational-choice theory as an analytical tool for explaining the economic and political behavior of individuals and nations. The theory is that actors will make personal choices in ways that maximize their self-interest. In economic terms, this generally refers to profits and economic advantage; in political terms, it usually means wealth and political power. Central to this concept is the idea of individualism; the cooperation necessary for living in a civilized society is overlooked. Some have suggested that even cooperation is based on individuals' cost-benefit calculus to maximize their self-interest. Much of this theory reflects not biological or psychological realities but the reality of a market with minimal state intervention. "Rationality" in the Cuban context, in contrast, is not necessarily linked to or derived from economics but defines the human-to-environment relationship in terms of development and social justice.

I now turn to the thoughts and analyses of a selection of Cuban intellectuals and officials to try to distill a greater understanding of the intellectual currents circulating on the island in relation to the concept of sustainable development. The analyses set forth by the following Cubans demonstrate a very thoughtful and often critical analysis of not only the concept of sustainable development, but also of the way it has been interpreted internationally as well as the obstacles Cubans and other societies face when trying to make sustainable development a reality.[1] From examining these ideas, we can gain a deeper understanding of how the concept has taken shape on the island and how the Cubans themselves view the concept as well as how their nation is trying to implement the principles set forth in Agenda 21.

CONTEMPORARY CUBAN THINKING ON SUSTAINABLE DEVELOPMENT

When Cubans view development through a historical lens, they tend to point to the impact of the Spanish colonizers in implanting a pre-capitalist system that evolved into a capitalist and eventually a socialist one shaped by socialist humanism and central economic planning. The Cuban academic José Mateo Rodríguez sees the capitalist development model as imitative and consumerist, driven by transnational corporations and resulting in an economic system with large external debt and small pockets of concentrated capital, and economic marginalization for the rest of society. Economic growth depends on an increase in production, leading to maximum utilization of resources, and environmental degradation. Mateo argues that Cuban development in the post-1959 era was austere, directed toward the satisfaction of the basic needs of the population, and carried out in an "atmosphere of equality, equity, and homogenization of all the social sectors" (Mateo 1998: 3). The development model resembled that of Cuba's Latin American neighbors in that it resulted in some level of economic dependence, in this case on the socialist bloc. The production processes utilized by Cuba from 1959 until the 1990s employed imported technologies that required large amounts of fossil fuels. The environmental costs on the parts of companies were not incorporated into the productive process, and this contributed to the aggravation of environmental problems (Mateo 1998: 4). Especially in large cities, Cubans were "fundamentally consumers, who produced little than what was consumed, saved little, recycled almost nothing, and were subordinated to the logic of vehicular transportation." While production was organized collectively, they were in essence estranged from the means of production, the natural resource base. Cuba continues to face this dilemma, perhaps more acutely today. The opening to foreign investment, the introduction of market mechanisms, and the expansion of the tourism industry have opened a Pandora's box of consumption. According to Mateo, Cuba faces a great challenge in winning the hearts and minds of the people, for the new model of sustainable development.

Mateo identifies the balance between creating a reasonable humanist standard of living for people with a conscientious consideration of the ecological limits and equilibrium of nature as sustainable development: "One of the challenges is incorporating a perspective of sustainability into the process . . . in inculcating a real ethic of sustainability and an environmental culture, that can be used as a productive potential, an existential potential and even as a spiritual or ethical development potential for society" (Mateo 1995 cited in Mateo 1998). He asserts that Cubans have a great potential for achieving sustainable development given social advances of the revolution. Cubans are highly educated and have high levels of scientific and artistic development, a high-quality educational infrastructure, and relevant social capital, all of which he views as "palpable strengths" (Mateo 1998: 9). The greater strength of the citizenry, however, is its "solidarity and equity ethics, the imaginary collective centered in the patriotism developed in the Cuban people by the revolution" (Mateo 1998: 9).

"Cuban thought from Felix Varela to the national hero José Martí has . . . a consciousness, symbolism, and spirituality based [in part] on nature" (Mateo 1998: 8). Cubans from the nationalist struggle for independence to the revolution viewed nature as the means for the construction of a new society and nation. Whereas for Spanish colonizers nature was to be mastered for gain, and slaves and salaried workers were estranged from the natural means of production, in Cuba's socialist society, it was urbanization after 1959 that contributed to their estrangement from nature. While the difference between the two models is more than semantic—distribution of goods among the populace versus capitalist maximization of profit—the risk in the Cuban case is a fetishization of commodities.[2] In the past Cuba tried to avoid this by encouraging a certain austerity. Interestingly, this lack of access to consumer goods is often what many immigrant Cubans—particularly the émigré economists—hold up as evidence of both Castro's so-called mismanagement of the economy and unsatisfactory Cuban way of life. Perhaps they understand all too well that materialism and consumerism are necessary features of capitalism.

Cubans argue that education can counter the consumerism that is slowly seeping into the consciousness. Teresa Borges has argued that the fundamental cause of environmental problems is the insufficient conscience, knowledge, and environmental education (Borges Hernandez and Díaz 1997).

Many Cubans link their ability to move toward sustainable development directly to their advances toward social justice. Following satisfaction of basic foodstuffs, they point to the country's literacy (nearly 96 percent), its free education system (primary through post-secondary/technical), a free health care system that is considered one of the best in the Global South, and a commitment to social development and equity that plays a large role in measurable social development results as illustrated by the chart below. Despite its tiny per capita GDP, Cuba in 1999 rivaled the more "developed" neighbors to the north, Canada and the U.S., in all standards of living and notably outperformed them in terms of physicians per person (Table 9.1).

Access to prenatal care and birth control ensures that children are cared for before they enter the world and welcomed by their parents and those parents who have the ability to care for them. Although not frequently discussed directly by Cubans, birth control is an important factor in the country's social progress. As many nations of the Global South know all too well, large numbers of people without the means to provide for themselves and without a government cradle-to-grave social safety net are subject among other things to malnutrition, infant mortality, and child abandonment.

One of the most difficult dilemmas, which face so-called developing nations, is the demographic trap. In addition to increasing life expectancy lev-

Table 9.1. Selected Health Indicators, 1999

Indicator	*Cuba*	*Canada*	*Costa Rica*	*Dominican Republic*	*United States*
Per capita gross domestic product $US	2,208	20,822	2,942	2,091	32,778
Infant mortality rate (per thousand children)	7	6	12	44	7
Under 5 mortality rate	8	6	14	49	8
Life expectancy at birth (in years)	75.7	79	76	70.6	76.7
Physicians (per one thousand persons)	5.18	2.1	.85	1.5	2.6

Sources: Miren Uriate, *Cuba: Social Policy at the Crossroads: Maintaining Priorities, Transforming Practice*, Oxfam America, p. 9, drawn from United Nations Development Program 1999.

els, drops in infant mortality, access to potable water, and higher levels of nutrition, populations tend to soar. Changing the cultural norm of having large families, so as to offset the deaths of children and lack of birth control and family planning measures, can also be assisted by social security programs. In addition to raising the standard of living and providing health care and family planning, Cuba has addressed problems of caloric intake, social security, and social assistance. It sponsors a universal national pension program to which employers contribute 10 percent of their income. The retirement age for women is fifty-five and sixty for men. The safety net also includes disability insurance for workers and those involved in accidents, means-tested cash subsidies and aid for families with no economic support, and a universal food subsidy through the ration program that covered roughly 65 percent of a family's needs in 2000 (Uriate 2002: 7). Elderly people, children, and pregnant women are given extra rations of milk and other goods in addition to the traditional offerings. Cubans also receive assistance in the form of rent control (rent is capped at 10 percent of income) and housing subsidies; many Cubans have been able to purchase apartments from the government and thus enjoy the benefits of home ownership. They also receive utility subsidies from the state for gas, water, electricity, and telephone service.

Cubans link the progress in social development directly to the ability to pursue a more sustainable society. They consider education a right and so it is free from primary to post-secondary. University education is free for those chosen by competition based on ability and economic needs. Students who do not qualify at first academically are able to join a community service organization and following their service (for which they receive a small salary) they earn credits to facilitate their entrance into a university program. Health care is also a right and offered without cost to Cubans. Through the decentralized *policlinico* system of local-based health centers and a focus on preventive care, Cubans have health care service in proximity to their homes.

CUBAN VIEWS ON SUSTAINABLE DEVELOPMENT

We see concepts and ideas on sustainable development throughout the writings and views expressed by a variety of Cuban contemporary thinkers. There are two primary themes in these thoughts—the practical and the theoretical. We will discuss the practical first along with its three sub-themes around which we can organize these perspectives: first is the context—both global and domestic—in which Cuba's transition to sustainable development is occurring; second is the relationship between Cuban citizens and Cuban citizens and the state; and third is the self-conscious thoughtful critiques about the process of the transition to sustainable society in Cuba. The second theme centers on the relation between humans and nature within a Cuban perspective.

GLOBAL AND DOMESTIC CONTEXT

Ricardo Sánchez, vice-minister of the Ministry of Science, Technology and Environment, notes the importance of the Rio Declaration as a necessary step toward sustainability and stipulates "consumption patterns and production patterns in the world have to change, otherwise there will be no sustainable development" (Sánchez 1998). Sánchez notes this is the path Cuba has taken since signing the accord in 1992 and especially since the collapse of the socialistic bloc.

Silvio Baró Herrera differentiates sustainable development from current neoliberal economic and globalization policies, the latter of which he views as irrational, unjust, and ethically unacceptable. Instead he views a campaign to develop a culture of sustainable development to facilitate consumption patterns suited to the context of a given nation a necessary precondition for sustainability: "Sustainable development urgently demands development of a new culture by way of models . . . styles of development and consumption" (Baró 1998). But he asks whether this objective can be reached under a globalized neoliberal system where "the main world powers have control over the media and maintain propaganda campaigns that are not limited to just promoting the consumption of goods aimed at satisfying real needs, but some are also aimed at creating false needs . . . linked to social status" (Baró 1998).

Juan Antonio Blanco Gil, founding president of the Felix Varela Center of Cuba, has identified threats posed by the spread of transnational finance capital to the construction of a sustainable society. "The danger now threatening humankind is not technological globalization, but the totalitarian transnationalization of financial capitalism under the principle of maximizing profits by passing the costs on to the environment and to human beings" (Blanco 1998). Blanco points to the need to rethink the relationship between the state, the market, and civil society. He points to the need for participation in the construction of a sustainable society. "This democratic level of communication, understanding and cooperation is also an absolutely necessary component of a sustainable society and of a sustainable international community" (Blanco 1998).

Gloria Gómez País questions whether any Latin American nation can create a sustainable society within an international context of globalization and neoliberalism and asks what practical meaning concepts like global warming and biodiversity can have for people simply struggling to survive (Gómez 1998). She argues that the relationship between poverty and environmental degradation must be addressed before a transition to sustainability can occur. She situates the challenge of creating a sustainable society within the reality of an international order driven by globalization and informed by models of neoliberal economic development. Gómez views Agenda 21 not as an environmental agenda but fundamentally as a development agenda with a political program.

Gómez notes the importance of solving problems of poverty and meeting basic needs as well as the necessity of environmental education. She points to the necessity for the transfer of cleaner production technologies, most of which have been produced in the Global North, to the Global South. But because of the reality that given the high costs of the technology and the unwillingness of the North to affordably transfer these advances to the Global South, another route must be found (Gómez 1998). Efforts between nations in the Global South to share technologies that they themselves have created are important steps in that direction.

José Ramón Fabelo argues that we cannot only view socialism from an economic point of view but also from both cultural and social points of view (Fabelo 1998). He argues, "humanity needs a new ecological and socially responsible culture that the market, by itself, cannot guarantee" (Fabelo 1998).

CUBAN CITIZENS, CUBAN STATE

Analyzing the Cuban culture in terms of crafting a sustainable society, Jorge Nuñez Jover, a sociology professor at the University of Havana, points to the role of Cuban scientists with regard to the changes that took place in the 1980s and 1990s and the role of socialism in the overall development program and argues: "The task of development has to be a process that mobilizes social creativity both to identify its objectives and to struggle for them. The struggle for development, above all, is a hallmark of socialism, demanding generalized social innovation directed toward transforming institutions and social practices in accordance with planned objectives. Development, equity, justice, and co-responsibility cannot come without a significant national capacity to produce, disseminate, and apply knowledge" (Nuñez 1999: 205). This of course, is founded upon the attainment of social progress and development as witnessed by Cuba's achievements in literacy, education, health care, and standard of living.

PERCEIVED CHALLENGES TO THE NEW MODEL

Cubans point to the difficulties they face in building a sustainable society in the current international context, and are quite forthcoming and thoughtfully self-critical in examining their own roles and that of their nation in this process. For example, the urbanist Gina Rey calls attention to the fact that in the globalized world economy, "more than ever the survival of the fittest prevails in the economic, political, and ideological realms." In her view, in contrast, economic growth "must unquestionably be well-being, which means not only material wealth but independence, social justice, human solidarity, citizen security, and 'veneration of full human dignity' as José Martí put it" (Rey 1998).

Ania Bustio argues in the Cuban case the variables of economy, society, and the environment are dialectically interconnected. Cuba has placed humans and their basic needs at the center of the analysis to development (Bustio 1998). Further, there is a realization that protection of the environment is necessary to create the conditions for a "rational life" for the present generation and ensure the survival and well-being of future generations. He lists a number of points that Cuba must attend to on the road to sustainability: defending the revolution and its gains, meeting people's basic needs, cultural, material, and the environmental, developing the environmental consciousness of the people to solve the current environmental problems and avoid the creation of new ones, strengthening environmental principles in national, regional, and local policy, and applying the concept of sustainable use to the biological resources because they are the foundations of Cuban industry (Bustio 1998: 8). Bustio argues that Cubans must examine the external and internal factors that can hinder sustainable development. One must, of course, be cognizant of the impact of the economic blockade and other actions by Cuba's neighbor to the north. Other internal factors, over which Cuba obviously has more control, include insufficient economic structures due to centralization and inefficient handling of natural resources, the limited capacity for environmental management and the limited degree of environmental consciousness in organizations and the public, excessive urbanization, unsustainable methods of resource use based on inadequate distribution of technological advances, and the insufficient development of environmental ethics through formal and informal education (Bustio 1998).

Additionally, both Bustio and Rey stress participation. The 1997 framework law calls for public participation in the policy implementation process, and the participation of mass organizations and nongovernmental organizations is an important start. Environmental education is a necessary precondition for such participation just as literacy is prerequisite for full political participation in governance. It is imperative that Cuba stay on course and not be afraid to more fully incorporate public participation in such decision making. In addition, as Cubans have shown, environmental education is a key ingredient in decision making, but the development of environmental ethics regarding human relation to the natural world is also a necessary aspect.

HUMANS AND NATURE

Manuel Iturralde-Vinent, a professor of geological science and researcher at the Natural History Museum, stresses the importance of education in the protection of nature: "When economists and business people alike are raised in an environment in which nature is an essential category; when decision-making persons know for a fact the irreparable consequences of their actions, when the ethics of these ideas are embodied by all and not the criteria of a few, then, we will be more cautious in our pronouncements toward the natural environment. . . . To paraphrase the poet, only love generates wonderment and only love perpetuates it" (Iturralde-Vinent 2002: 8).

Urban planner Gina Rey also recognizes the connections between ethics and morality in the construction of sustainable development. She reasons there is a decline not only in the quality of life for many residents of Latin America and the Caribbean but also a "deep degradation in people's moral and ethical values" (Rey 1998). Rey points to linkages between resources as well as the conceptualization of humans' relation to the natural world in the development questions. She notes unjust social structures, wealth concentration, and unsustainable development practices as well as political systems, which do not include public participation, cause environmental degradation and must be changed" (Rey 1998). She notes a change in the mindset is required which not only conceptualizes the human-nature relationship in a new way, but also requires true participatory democratic processes. She says the existing globalized economic system is comprised of "irrational societies that have contaminated and destroyed nature, while concentrating wealth in a few hands in which energy is wasted, consumerism and individualism go to excess, all of it ultimately to bring about not well-being but alienation and loss of the essential values of the human being. At least now it is known that this is the antithesis of the development model that humankind needs" (Rey 1998).

Olga Fernández and José Fabelo argue that an anti-anthropocentric approach is not what is necessary to resolve the ethical question over human impact on the environment in terms of sustainable development (Fernández and Fabelo 1998). Instead they argue it is not the focus of our analysis (human centered or anthropocentric or another), which must change, but humans themselves as well as the social and economic contexts in which we exist. Humanity, whether in the Industrialized North or "developing" South, needs to come to terms with the fact that the natural world has limits and so the principle of maximum gain as the driving force of development must change (Fernández and Fabelo 1998). They argue we need a new ecologically and socially responsible culture, which prioritizes social justice, environmental protection, and humanism.

Further Fernández and Fabelo point to the need to focus society on *true needs* not in the sense of traditional models of consumption, but what humanity's true needs are. Bringing in an analysis of consumption to the questions of development based on social justice concerns as well as realization of human potential will help bring us closer to a sustainable model of development.

TOWARD A "CULTURE OF NATURE"

One of the most important Cuban thinkers on the relationship between humans and nature was the late Antonio Núñez Jiménez, who played a particularly interesting role in Cuba's revolutionary transition, first as a student activist and then as a captain under Che Guevara and eventually head of the agrarian land reform institute and first president of the National Academy of Sciences.

In 1994, he created the Foundation for Nature and Man, a Cuban environmental NGO, and in his address on that occasion, he said, "This organization is born in order to struggle for ecological values, to improve the health of a planet sick with contamination. . . . And it is also born to struggle against the contamination of humanity and against the political plagues which in many parts of the world aim to turn humans into third-rate pariahs. . . . We struggle, in other words, for a happier humanity living as part of a healthy nature" (Núñez Jiménez 1999: Introduction).[3] It was Núñez Jiménez who coined the phrase "toward a culture of nature" and embarked upon a campaign of the same name in 1980 that coincided with the promotion of a popular camping program in Cuba, presumably on the assumption (similar to North American John Muir's) that if people experience wilderness they will be more likely to protect nature (Núñez Jiménez 1999: 1, Muir). Núñez Jiménez's campaign and camping program were aimed at raising awareness regarding environmental problems and crafting a respect for nature as part of a Cuban identity (Núñez Jiménez 1999: 1).

Núñez Jiménez saw the importance of relating Cuba's past and present as well as its place in the world to the connection between humans and their place in the world. In the introduction to his book, *Hacia una cultura de la naturaleza*, he expressed these sentiments as follows:

> Neither philosophy nor ecology is possible, without a scientific point of view, to deal with nature and the environment of Cuba. . . . We live on a single earth. . . . We breathe a single atmosphere. . . . We are surrounded by a single ocean, and politically we live in a world, in which, though divided into nations, rich and poor, developed and underdeveloped (an abstract euphemism for exploited and exploiter), [where there is] contamination of the Ocean, in the air we breath or the earth, therefore, we are all affected. An atomic blast directly or indirectly in one corner affects the air, the ground and the soil of all nations, and this can also be said of the industrial pollution or oil spills.

This comprehensive volume on the development of an environmental consciousness is a tour de force in the history of environmental thought, and it places Cuba in the context of its global examination of environmental issues and ideas.

Núñez Jiménez's long-time friend, Angel Graña González, general coordinator at the Foundation for Nature and Man, reminisced with me in 2000 about his impact on Cubans' perceptions toward nature (Graña 2000). When he began his campaign in 1980, Graña said, there was no environmental culture in the country. From that time on Núñez Jiménez had worked to develop environmental consciousness in ordinary Cubans. He produced a regular column in the party newspaper *Granma* called *Man and Nature*, and in it, he taught Cubans about the geography of their country and shared stories of his many adventures exploring the planet. Further, he helped to bring about what became known as the SOS Round Tables, which continue to bring together stakeholders from the area as well as scientists, technicians, and government representatives to discuss particular environmental problems. The foundation has held discussions on pollution in Havana Bay and on the idea of damming the Río Toa for a large hydroelectric project. The Toa River valley contains the only lowland rainforest area on the island and is considered equal in terms of biodiversity to Amazonia (Graña 2000). The participants argued that the sediment runoff from the dam project would eventually destroy the rainforest and the project was halted, in particular response to the efforts of Núñez Jiménez.

According to Graña, Núñez Jiménez influenced those in government as well: "There is no doubt he was the main figure in environmental protection. He taught the government the need to take care of the environment—that the responsibility of this generation [is] to preserve the environment for future generations." The special influence he wielded, however, was not simply due to his being part of the revolutionary government. Throughout his books, particularly *Hacia una cultura de la naturaleza,* are scattered photographs of Núñez Jiménez with Fidel, Raúl, and Che, particularly when they were hiding in the Sierra Maestra mountains. Below the caption of one such photograph, he wrote, "Healthy and fruitful contact with nature will lead to an improvement in the natural resources of Cuba (Núñez Jiménez, 1998: 515). While Núñez Jiménez was sowing the seeds of Cuba's revolution, it appears that Cuba was sowing its own seeds of an appreciation of nature. The memories of this time never left the man or Fidel. Graña attributes Núñez Jiménez's ability to influence Cuba's policies toward the environment to shared times and life-long friendship: "He had the ear of Fidel because they were friends" (Graña 2000).

In a column on air pollution in February 1980, Núñez Jiménez discussed the role of environmental protection in a nation that was both "underdeveloped" and revolutionary. He pointed out that people were beginning to be aware that "the traditional irrational use of natural resources, and the anarchy of the free market, particularly the industrial sector," threatened the current standard of living and long-term viability of nature for future generations. Here he connected all of the points that contemporary Cubans are making about sustainable development in their country.

He went on to target capitalism, "with its great industrial and technological development" as using natural resources in a "devastating and irrational manner," and blamed the capitalist countries, with their focus on private property and unplanned economies, not only for devastating their own natural resources but also for spreading technologies that damage the health of humans and animals and for the degradation of ecosystems.

Again, these sentiments are echoed among Cubans today.

Núñez Jiménez saw his "culture of nature" having two harmonious and parallel components: "love for and aesthetic sensitivity toward Cuban nature, its landscape, and its flora and its fauna and the best use or transformation of it so as to preserve it for the benefit of future generations, to which we must bequeath a more beautiful and useful Cuba" (Núñez Jiménez 1980). This is an exceedingly important linkage. The conjoining of ethics to an aesthetic of nature stemming from a love of nature was also discussed by North American Aldo Leopold: "To cultivate in the public a refined taste in natural objects is vital to enlightened democratic land use decisions" (Leopold 1953: 149).[4] The impulse to protect comes not from duty, which is perceived as a "limitation on freedom in the struggle for existence," but from love. We will protect the things we care about not out of duty, but out of love.

One must weigh questions of utility, duty, and beauty. "Contemplating human action in respect to land, one might suppose that utilitarian desiderata tempered and occasionally opposed by ethical criteria are preeminently efficacious while aesthetical considerations carry less weight," explained philosopher J. Baird Callicott in examining this conflict in Leopold's work (Callicott 1987: 158.) A situation of need such as that of Cuba in the Special Period might justify a utilitarian perspective whereas one based on ethics would be seen as necessary but more difficult to follow; however, love helps make that leap easier. By encouraging Cubans to experience wilderness, Núñez Jiménez hoped to foster a love of nature that would inspire Cubans to protect it.

Cubans have moved beyond a utilitarian, instrumentalist view of nature in the direction of protection of the natural world via an increase in knowledge about the human impact on the environment and an understanding of natural processes. The connections they see between their commitments to social justice and the societal focus on realizing human potential and the environmental ethics and aesthetics advocated by Núñez Jiménez and other Cuban thinkers is a substantive contribution to international environmental thought.

As this chapter has shown, Cubans conceptualize sustainable development as the joining of social justice and environmental protection, and inherent in that perspective is humans' relationship to one another as well as to the natural world. The aforementioned thinkers view a high level of social development as a necessary prerequisite to social justice in the society, and as we have seen in this case and within the literature, poverty leads to environmental degradation. But we also recognize, as do the Cubans, that as a society develops and standards of living rise, so does resource use. As Cuba opened itself to foreign investment and the resultant economic growth so came increased resource use and environmental degradation both in the economic development itself and higher levels of income, spurring the creation of the environmental impact regimen as well as discussions on questions of consumption and materialism. Mateo points to this and calls for a model of sustainable development that balances a humanist standard of living with recognition of living within the ecological limits of nature. Education is a necessary component for adoption of the humanist model as opposed to the consumption-oriented mindset, as Borges points out.

The development of the whole person as Rey, Fernandez, and Fabelo argue is necessary as opposed to economic development for consumption based on false needs. This is an essential lesson from socialist thought and the socialist experience of the Cubans that contributes to the understanding of sustainable development, that is, to organize society around the development of the whole person within the limits of nature as opposed to economically developing the society through the creation of false needs in people. The focus then is on both the basic needs for living and the fulfillment of needs for self-actualization that will lead to a model of sustainable development within the confines of nature—"a rational use of resources" according to the Cubans—because these needs have a point of satiety whereas false needs cannot be fully satisfied within the limits of nature. These lessons are important for any society grappling with the balance of human needs within the limits of the natural world. Both Núñez Jiménez and Mateo place sustainable development within socialism. Quoting Fidel Castro, "Development is in essence a continuity strategy, insomuch as idea of sustainability is intrinsic to the socialist principles that sustain our revolutionary model."[5] Indeed, as Mateo and others have shown, Castro's thoughts on the environment have evolved over time as well.

Núñez Jiménez makes an exceedingly important contribution to the Cuban discussion of sustainable development as well as to environmental ethics globally with his call for the need of a "culture of nature." He offers a guide to the development of an environmental aesthetic for others in his writings in raising the Cuban consciousness about their own environmental patrimony as well as the environmental wonders of the world around them. By educating average Cubans in his weekly newspaper columns, Núñez Jiménez fostered a Cuban appreciation of nature, and through his campaign on camping, he hoped to inspire a love of nature through the wilderness experience so that they might in turn protect the environment. For other Cubans who might not be able to participate in camping, Jorge Ramón Cuevas took Cubans along with him on his explorations of the natural world in his television program, *Entorno*. Cuevas also attempted to cultivate a love of nature in his audience. Both men revealed the wonders of nature and allowed Cubans themselves and make the emotional leap to the valuation of nature to hopefully appreciation and protection. The organizations that both men founded continue these efforts today with programs of environmental education for students as well as the general public in addition to grassroots environmental protection projects.

Taken together, the foci on the social, cultural, and creative development of humans as well as the cultivation of an environmental aesthetic and ethic situates humans within their proper place in the natural world and is a necessary step toward the creation of a sustainable society.

NOTES

1. This selection is not meant to represent an all-inclusive survey of intellectuals in Cuba, but is a sample of intellectual currents drawn from their own writings and conference presentations.
2. I am indebted to the late John L. Stanley for our discussions on the point of the fetishization of commodities.
3. Núñez Jiménez cited in Velis de Núñez Jiménez (current FANJNH president), June 1999.
4. Leopold, of course, draws upon the Platonic traditions as set forth in the *Symposium*. Leopold saw the relationship between beauty and love or *eros*.
5. Mateo 1998, quoting Fidel Castro, "Intervención ante la Asamblea Nacional del Poder Popular," Granma, Havana: Cuba, July 21, 1998, p. 4.

10

Lessons

> *We need to build up societies that are ecologically sustainable, that are human centered, that are socially responsible—societies that consider human beings not as consumers but as individuals who deserve to have their basic human needs covered. We are still trying to make it viable in Cuba in the middle of a war.*
> —Juan Antonio Blanco

Cuba's sustainable development policy emerged from a convergence of variables at a specific time and place during a political and economic crisis. It evolved from the research and efforts of scientists and technicians on sustainable methods of production, a societal commitment to social justice, and a government's need to survive against all odds to preserve the social-justice gains of the 1959 revolution. The case of Cuba demonstrates an example of operationalizing the concept of sustainable development, an aspect that many scholars in the literature have noted was needed to illustrate what the concept can mean for nations. This critical case study helps in evaluation of other cases in Latin America and possibly throughout the Global South. Cuba's transition to a more sustainable society holds important lessons for other nations in terms of what can be done when the political will to protect human and nature is present.

This chapter briefly reviews the lessons learned from this case, discusses the progress Cuba has made, analyzes and critiques the pace of the transition, and further draws out the lessons that might be applied to other cases in Latin America.

LESSONS LEARNED

Cuba, since the economic blockade was initiated by the United States, has had to seek some levels of self-sufficiency. As the island's relationship with the former Soviet Union warmed, the focus shifted somewhat away from self-sufficiency and more toward integration within the Council of Mutual Economic Assistance or CMEA. However, Cuba was not unaware and unaffected by outside larger trends going on in the rest of the world. For example, much of the agricultural technology it embraced was in the Green Revolution-style of the West although this technology came to Cuba via the socialist bloc. Cuba also felt the impact of the alternative agriculture movement that occurred elsewhere in the 1960s and 1970s and in a similar fashion to other areas of the world; these movements impacted a small group of growers but clearly were not the dominant approach. Cuba also felt the impact of the energy crisis that dominated the globe in the 1970s. Similarly to other nations of the world, these events had an impact on a small group but again were not the dominant approach. Research into these areas—agricultural and renewable energy resources—began in Cuba but did not come into play in a large way until the island found itself in a full-blown economic and political crisis with the collapse of the socialist bloc.

The rethinking that was going on with in the socialist bloc, *perestroika* and *glasnost*, had its equivalent in the Cuban case, as witnessed by the efforts of the Rectification of Errors and Negative Tendencies, or Rectification Process. At the same time, Cuba was in the midst of a Cold War-era defense to the cowboy adventurism of the United States under the lead of a "Cold Warrior" U.S. President Ronald Reagan. Cuban President Fidel Castro was also quite critical of the changes toward rethinking socialism and economic and political liberalization that were occurring in the Soviet Union under Mikhail Gorbachev. (While analysis of Cuba's Rectification Process is beyond the scope of this book, others such as Azicri, Eckstein, and Lutjens have examined the process in a thorough fashion.)

The Rectification Process and the Special Period in Peacetime—which followed the collapse of the socialist bloc—led to development of the food program, which began in 1991 and focused on food security. Carmen Diana Deere and others argued that in trying to reach the goals of food security, much experimentation occurred because the "central planning apparatus has found it increasingly difficult to plan, opening up the possibility for decision-making to become more decentralized. Moreover, differing local conditions have prompted different responses to the crisis, requiring flexibility in the implementation of national policies" (Deere et al. 1994: 195). This decentralization was attributed to the experimentation that Deere et al. argued contributed to the modest degree of success by the food program by 1994. Indeed, as we look back a little more than a decade later, we can argue the decentralization and experimentation has not only contributed to the great success of the food program in all its new forms, but also enabled it to grow and expand. The decentralization and experimentation has also played a role in the formation of the nation's development of local-based self-sufficiency in agricultural technology seen in the formation of the Centers for the Production of Entomophages and Entomopathogens, or CREEs (which produce biological control agents), as well as in the development of the expanding renewable resource industry for domestic use and potential future export.

Deere, Peréz, and Gonzales argued the change of land ownership seen in the development of the UBPC (Basic Units of Cooperative Production), as well as the more recent developments of urban agriculture and *organipónico* system, have expanded exponentially and played an enormous role in the decentralization of Cuban agriculture. They concluded, "while recent developments suggest that the Cuban government is moving away from the previous state-centered model of capital accumulation, we argue that what is emerging is far from a 'restoration of capitalism' or a privatization; rather, some promising preconditions for socialist (or communal) class relations are developing in their period of transition, including the break-up of many state farms and their conversions into structure resembling production cooperatives and which are characterized by worker participation in decision-making and self-management" (Deere, Peréz, and Gonzaléz 1994).

The continuation of these agricultural relations as well as experimentation in agriculture and renewable resource development in the last decade again demonstrate how the decentralization and experimentation have taken on a life of their own and made much progress toward socialist or communal class relations in these two areas. We also see this type of experimentation and decentralization occurring within government itself in the body of state agencies. Many state agencies, like the CREEs, agencies within the Ministry of Agriculture, and programs within the University of Havana are being told they must become financially self-sufficient. Thus many experiments and decentralization exercises are resulting in a wide-ranging decentralization of the state.

Sheryl Lutjens describes the dynamic between participation and administration of the state in the case of Cuban education in her work, *The State, Bureaucracy and the Cuban Schools*. "Cuba's survival strategy preserved the promise of decentralization and the revolution's exceptional devotion to education," wrote Lutjens (Lutjens 1996).

At the same time, there are interesting trends occurring within the Cuban economy since the passage of the foreign investment law in 1995. The Cuban government has entered into a variety of pacts with foreign investors in creating co-owned enterprises, with the Cuban government usually, but not always, owning the controlling interest in the business. While certain social service-related industries are not permitted to be owned by foreign investors, two of Cuba's largest industries, tourism and mining (and the most potentially environmentally destructive), fall within the public-foreign ownership sector.

While these two are seemingly disparate developments of decentralization and foreign investment that characterize the island since the collapse of the socialist bloc, there is a clear link, which binds them. The Cuban government has embarked on a journey toward a more decentralized state and mixed-economic market through the efforts to manage and operate their businesses, participate more fully in education, and slowly decentralize portions of the state bureaucracy. The government has opened some sectors to foreign ownership or shared ownership, usually controlled by a larger state share of 51 percent, while other sectors of the economy are off-limits to foreign owners. The move opens the country to foreign investment, the global market, and possibly toward a mixed economy. At times, ordinary Cubans are also permitted to participate with these efforts as seen in the *organipónicos,* development of small Cuban-owned businesses like restaurants—*paladares,* lodging—*casa particulares,* even small fast food centers and small merchandise stands. A variety of quasi-legal or even black market services like baking pastries, video rental, cueing up for others in the *libreta* (ration) line, and other activities of entrepreneurship go through periods of tolerance and intolerance by Cuban officials.

Fundamentally, the Cuban state sees the necessity of preserving the gains of the revolution in social and economic development. Gradually there was a realization that the bureaucratization of the state had to be diminished, but since the state was not withering away on its own so to speak, active changes were necessary. A precondition to this change necessitated the fuller empowerment of the people. Clearly in the early years of the revolution, Cubans needed social and economic development ranging from ending hunger to learning to read. For the revolutionary Cuban state—whether in its current form or another form—to see the gains of the revolution survive, within the people themselves, the people then had to be empowered separately from the state. This helps explain the moves toward worker self-management and decision making that we have seen in agriculture and throughout the society. But this is not a smooth or even consistent progression. The process moves by fits and starts. The situation is further complicated by the introduction of foreign participation of outside investors as well as the ever-changing geo-political dynamics, particularly with its powerful neighbor to the north, but also relationships with Western Europe and Canada.

LESSONS FROM THE CUBAN EXPERIENCE

Cuba also has maintained its commitment to social development and equality within the society in addition to environmental protection. As the

Cuban experience shows us, without literacy efforts to educate people regarding natural resource protection, these programs could not succeed. Education and social development go hand in hand in helping facilitate a people's ability to look beyond their day-to-day survival to future generations and making small sacrifices so as to not endanger intergenerational equity, that is, protecting the future generation's ability to survive. Additionally, without access to renewable resources, people will turn to a country's natural resources for heating, cooking fuel, and the like to enable them to survive and in the process degrade the environment.

Birth control and family planning allow citizens to make a choice about when they are financially, physically, and mentally able to successfully care for and raise a child while not endangering the parents' and other siblings' ability to survive. Access to birth control and pre-natal, maternal, and infant care is necessary as are social security measures so children do not become, as is the case in many parts of the Global South, the sole pension packages for their parents. Another essential precondition for this is a drop in the infant mortality rate. Birth control and family planning are essential to any nation, particularly those attempting to overcome the demographic trap of development—in that without increased populations states are more able to care for the existing citizens in terms of infrastructure like housing, access to potable water, food, sewage, education, health care, employment, social security, and the like. Population pressures must be contained in industrialized and so-called developing nations so that the standard of living can be increased within the limits of the natural resource base on which human life depends regardless of the level of sustainable technologies employed. Resource use, of course, must also be addressed, in particular by the industrialized nations.

The gains in literacy, education, health care, infant mortality, life span, as well as ending hunger are necessary preconditions for sustainable development. They comprise the social justice aspect of the approach along with environmental protection and utilization of sustainable technologies.

"In order that this transition can assure the proposed goals of justice, solidarity and equity, the objective of national independence must be permanently affirmed and underdevelopment, overcome," writes Cuban sociologist Jorge Nuñez (Nuñez 1999: 205). This transition was also carried out during an exceptionally difficult period—socially, politically, and economically for Cuba—following the collapse of the socialist bloc and in the face of the enhanced hegemonic power in the United States, which defined among its foreign policy proposals "the defeat of the Cuban revolution," argued Nuñez (Nuñez 1999: 205).

Cuba found itself in precarious conditions in the late 1980s and early 1990s, and clearly the island has not only weathered the storm following the collapse of the socialist bloc, but it has also prospered. Much of that prosperity came as a direct result of the successful implementation of sustainable models of agriculture, local food production, energy self-sufficiency, and attempts to monitor the environmental impact of development as foreign investors move to invest on the island. Throughout the process however, the fact remains that the transition was very much shaped by the Cuban commitment to social justice and equitable distribution of wealth among its citizens. That point is quite likely a *necessary condition* for a true transition to a sustainable society in Cuba and elsewhere. Also it places on the table the necessity for social justice and socially just development for humans and the natural world in the creation of sustainable societies. Moreover, it sets forth a number of challenges to other countries attempting to move toward a sustainable society.

When I first visited Cuba, it was in the early stages of implementing sustainable methods in agriculture, and I was told by some observers that these were only temporary measures—that when economic conditions improved and agricultural chemicals and pesticides began flowing into Cuba again, this greener model of agriculture would be abandoned. The scientists and technicians participating in the transition suggested that, given successful results, this approach could change, but in fact what made the pursuit of sustainable agriculture permanent was something deeper.

Cubans are committed to the survival of the revolution and the pursuit of social justice, and they know that survival is linked with sustainable development. They recognize the importance to this task to the development of a "culture of nature" and not only a well-rounded understanding of human impact on the environment but an environmental ethics instilled in education. Their efforts have earned them international acclaim and have been both a source of pride and the subject of thoughtful critique.

INTERNATIONAL RECOGNITION

The recognition granted Cuba by the United Nations agencies and international NGOs in recent years allows us to draw some conclusions about its progress in environmental protection and rehabilitation.

In October 1999, it was awarded the 1999 UNESCO Saving the Land award for its anti-erosion efforts in Guantánamo province (Radio Havana 1999). Officials of the Ministry of Science, Technology, and the Environment reported the recovery of more than three hundred hectares of land and the launching of an environmental education program throughout the country. In November 2000 these officials noted that pollution controls had resulted in a 10 percent reduction in water pollution in eight of Cuba's river basins (Radio Havana 2000). Improvement was particularly notable in the Cauto River basin, with a 7 percent reduction.

The Jose Martí Pioneer Children's Organization and CUBASOLAR, were named recipients of the United Nations Environmental Programme's (UNEP)

Global 500 Roll of Honor in May 2001 for their active role in environmental protection (Radio Havana 2001). The organizations were honored for their accomplishments in "the electrification of rural communities, the fight against desertification in the San Antonio del Sur Valley, the rehabilitation of Camilo Cienfuegos City School, the rehabilitation of the River Sagua Basin and the establishment of medical clinics and schools" (UNEP May 2001). The UN environmental agency noted that CUBASOLAR's efforts had greatly impacted the island in the areas of forest and ecosystem conservation, "recovery of saline soils and desert areas, the management of protected areas, the preservation of atmospheric resources, the enhancement of rural health and education systems, and the improvement of human settlements."

The organization was also lauded for its public education efforts and for its coordination of education and technology transfer among related professionals.

Additionally, Havana was selected along with Turin, Italy, as a site for the World Environmental Day, June 5, 2002. In March 2002 a group of six University of Havana researchers received the Whitley Laing Foundation award, from a British NGO, for their work on bird protection (*Granma International* Mar. 2002). A project on rice plantations and wetlands as conservation units for aquatic birds in Cuba, headed by Lourdes Mugica, focused on seventy bird species, many of them migratory, that winter on the rice plantations of Sancti Spíritus. In a *Granma International* interview, Mugica pointed to the importance of a "strong environmental education campaign to convince campesinos of the need to protect these birds, which are part of the Cuban natural heritage" (*Granma International* Mar. 2002).

A wildlife refuge project under the leadership of Ariam Jiménez was recognized with an award from British Petroleum in April 2002 (*Granma International* May 2002). In September 2002 Cuba received a $30,000 award from the United Nations Development Program (UNDP) to continue its reforestation work along the Cauto River (*Granma International* Sept. 2002).

Cuba has also been lauded for its work on cleaning up Havana Bay, once considered one of the ten most polluted bays in the world (*InterPress News Service* Sept. 2000). Official figures from late 1999 indicated that the 5.2-square-kilometer bay contained some twenty thousand cubic meters of waste (*InterPress News Service* Sept. 2000). A task force was set up in 1998 to begin the cleanup effort. The island is working with representatives and donors from the UNDP, Germany, Italy, Japan, and Norway on both cleanup efforts and projects to reduce pollution from sewage and oil spills. In addition to the construction of sewage treatment plants along rivers that feed into the bay such as the Almandares and the Luyano, efforts are also being directed at more than one hundred industrial plants that routinely dump waste into the nation's waterways or the bay itself. Officials report that the cleanup will include efforts to oxygenate the water and reduce the sedimentation through companion reforestation efforts along the waterways that feed into the bay which should encourage the return of sea life and birds as well as reduce the stench. Similarly, cleanup efforts will extend to the large ships that visit Havana Bay, which frequently cause oil spills such as the 120-ton spill in 1998. The government has begun purchasing state-of-the-art containment booms and oil skimmers for oil cleanups (*Radio Havana* June 2001).

Cleanup efforts have also begun in Santiago Bay and include reforestation efforts along the waterways upstream as well as industrial retrofitting efforts and sewage treatment plants (*InterPress News Service* Sept. 2000). According to Maria Josefa Rodes, an environmental expert and task force member, the government is also offering economic incentives to industries to join the nation's environmental conservation efforts (*InterPress News Service* Sept. 2000, Radio Havana 2001). In June 2002 the UNDP and the Cuban government announced that fish and birds had returned to the bay—a sign that pollution levels have dropped—but cautioned that the environmental cleanup was not yet complete.

Also in that year, the UNDP program in Cuba received the Elena Gil Award from the Felix Varela Center for its role in the human development at the local level, especially concerning gender relations and the environment (*Radio Havana* June 2002). The project focused on local development—both economic and social—as well as environmental protection and the credit system for small and medium enterprises (R*adio Havana* June 2002).

The international community has also bestowed honors on Cuba for efforts in the areas of sustainable agriculture and urban agriculture. In 1996 the Cuban Organic Farmers Association received the Saard Mallinkrodt Award from a Danish organization for its work in organic agriculture, and in 1999 the group was granted the Alternative Nobel Prize or the Right Livelihood Award (Funes 2002: 14–15). The FAO has also praised Cuba in a 2003 report for its success in urban agriculture (National Information Agency Aug. 2003). Since 1989, Cuban use of pesticides has dropped twenty-fold and this has resulted in a drop in pesticide poisonings, and in the use of precious foreign capital to purchase chemical pesticides abroad, and generated new employment opportunities for Cubans (Pages 2004).

An address that was part of the closing ceremony of the International Conference on Environment and Development in June 2003, UNEP Regional Director Ricardo Sánchez called attention to Cuba's accomplishments and made a case for sustainable development in Latin America based on a number of aspects from the Cuban model (Sanchez 2003). For example, he pointed to the importance of land stewardship by farmers and the opening of agricultural markets for their organic produce. Sánchez also identified the role of urban agriculture in alleviating food security problems, enhancing employment opportunities, and reducing environmental contamination, drawing examples from the Cuban experience. He emphasized the importance of social development and social justice in the examples of reductions of infant mortality and for education to help young people attain their full potential, and advocated self-realization in contradistinction to consumption, which he termed the "poison of people" (Sanchez 2003).

INTERNAL ASSESSMENTS

The most recent report on the environment of the Ministry of Science, Technology, and the Environment, *Situació ambiental 2002*, released in late 2003, noted a continuation of efforts to reduce pollution, halt desertification, and cleanup the production process through reuse, reclamation, and recycling. The report called attention to progress in environmental education, especially through television programs such as *University for All* series and the *Round Table* series on current events. It acknowledges the need to continue to move to a more participatory form of environmental management. Challenges remain, however, for the construction of a sustainable society.

CHALLENGES

On December 25, 2004, President Fidel Castro announced to a closed session of the Cuban National Assembly that a cooperative effort with Canadian companies Sherritt and Perbican had discovered oil off the coast of Santa Cruz del Norte, east of Havana (Associated Press 2004, Canadian Press 2004).[1] The announcement was made to the Cuban people on television the next day. The discovery was estimated to have a potential yield of up to one hundred million barrels of oil. Since Cuba currently uses about 150,000 barrels of oil per day, the new source could theoretically provide the island with a little more than eighteen months of petroleum, if per diem consumption remained constant.

The island currently produces only about half of the oil it uses, importing much of the rest from Venezuela. Further early reports say that the oil is of a higher grade (i.e., less contaminated with other elements, such as sulphur, found in much of Cuba's oil), which will reduce the costs of the refining process (Associated Press 2004, Canadian Press 2004). This discovery could allow Cubans to disregard the lessons in the past fifteen-plus years and put off dealing with what is, according to many experts, an inevitable decline in oil supplies both domestic and global.

While the attraction is no doubt appealing to do as much of the world has done and ignore this impending reality, Cuba should keep an eye on the months and years ahead. Indeed in June 2005, Castro acknowledged that the world will soon face a grave energy shortage as well as environmental crisis brought on largely because of the burning of fossil fuels (AIN June 29, 2005). Most mainstream climate scientists agreed that the burning of fossil fuels has definitely contributed to global warming, and in 2005 scientists from the Scripps Institute of Oceanography, Lawrence Livermore Laboratories, the U.S. National Oceanic and Atmospheric Administration, and the Hadley Center found an "unequivocal link between man-made greenhouse gases and a dramatic heating of the Earth's oceans" (Connor 2005). Researchers reported a "stunning" correlation between human-induced atmospheric pollution and a rise in ocean temperatures over the past forty years. The researchers said the findings *end* the debate as to whether humans have played a role in creating global warming. Therefore, while Cuba now has more oil than in the recent past and will no doubt make use of it for internal consumption as well as to raise foreign capital, it should not turn away from the progress it has made. As residents of an island, Cubans are cognizant of the impact of global warming on ocean levels and the hydrological cycle. According to Jorge Luis Aspiolea, president of the National Institute of Hydraulic Resources, in 2004, Cuba experienced the lowest levels of recorded rainfall since 1901, less than 70 percent of the country's historic average (Pages Feb. 2005). It was required to spend some U.S.$20 million to offset the lack of rain, and Cubans point to global warming as an important cause for the drought. Global warming causes warming in bodies of water and the atmosphere resulting in a disruption of the hydrological cycle that can cause a reduction in rainfall. To cope with the drought, officials called for water conservation and repair of leaky fixtures in homes, industry, and water supply lines, but conservation alone cannot solve the water shortage. Therefore, the island must utilize this new supply of oil wisely, perhaps drawing upon the foreign capital from its sale to continue its commitment to development of the alternative energy technology or perhaps the construction of new public transportation systems within the cities as well as repair of its declining water and sewer systems to offset existing pollution problems. The Cubans might look to the city of Bogota for an interesting example of a more modern sustainable public transportation system. Further, the island would be much better served to focus on development of a sugarcane-based ethanol system, similar to Brazil's, for its domestic consumption and possibly exportation. As Cuba's energy independence grows, so will its security, economic, political, and environmental.

Clearly there are challenges ahead for Cuba particularly in the area of public participation. As discussed in chapter 8, public participation in environmental decision making and management must be strengthened in Cuba. My interviews with Cuban officials as well as the words of Cuban intellectuals highlight the need for environmental education to increase informed participation by the Cuban people in environmental decision making and management. Just as it was necessary in the early years for the society to feed its people and educate them so that they could see beyond day-to-day survival to build a better society, citizens now need to be educated as to the connections between human life and the natural world, or in other words, the human impact on the environment. Unless these ideas are made concrete in the experience of Cubans, young and old, environmental protection may not survive beyond the revolutionary government.

Popular participation needs to also occur with environmental projects. This does not mean opening doors for external forces to oppose the direction development is taking in Cuba, but meaningful local community participation in decision making seen in the example of local-based decision making of the Havana Metropolitan Park. This process needs to be strengthened and expanded upon as not opening the process up to people under the guise of "national security concerns" will rollback the environmental protection efforts the nation has so far achieved. Cuba is making an important necessary step in the pursuit and implementation of environmental education process.

The retooling of industry for a cleaner production process is identified as important in environmental ministry documents; the campaign soliciting worker participation in cleaner and more efficient production is an encouraging sign. Environment ministry officials have outlined a program of annual meetings with representatives of other government agencies and domestic and foreign industry to educate them on the environmental laws and environmental impact review and licensing programs (Borges 2003). These efforts need to be expanded so as to reduce the occurrence of violations and create a smoother environmental impact review process. It is imperative that in the face of economic difficulty the process not be hampered. Continued transparency on the part of government will enhance the likelihood that the process remains intact.

Additionally, challenges to environmental management frequently arise from contradictions in the economy itself, specifically in mining and tourism, though contradictions are not necessarily unique to Cuba. Though industry with heavy foreign investment and tourism threatens the island's progress in social and environmental progress, they appear to be necessary evils. Yet these gains should not be lost. Industry with heavy foreign corporate participation has been both a burden and a boon for Cuba. In some ways, it was that involvement that helped produce the country's sweeping environmental framework law, and as foreign investment increases, the state must continue to insist on adherence to its regulations. (However, the disparity between earnings in the tourist sector and earnings in industry or government is well documented. In addition to the resulting exposure to foreign television and goods, there is a devaluation of the social progress that Cuba has achieved when scientists, physicians, and other highly educated persons can make ten times their salaries by working as a taxi driver or tour guide.) Education has always been highly regarded in Cuba and a love of learning permeates the political culture. Given widespread recognition that education does not necessarily lead to a better life, at least in economic terms, in encouraging tourism Cuban officials may be facilitating the destruction of the social fabric they have worked so hard to help create. A change in societal values as a result of the growth of tourism could prove to be a slippery slope. Without public dialogue about the dangers of consumerism, the market mindset that has already taken root in some sectors may grow and flourish.

CONSUMPTION

I am *not* suggesting from a Global North privileged position that ownership of consumer items should be forgone for the sake of ideological purity or Cubans should be denied the right to possess the accouterments of modern life. My point is that possession of consumer goods beyond those required for the necessities of living and self-actualization may erode the values on which Cuba is based. In other words, there could be a reification of goods. In consumeristic societies, people begin to seek and falsely find psychological fulfillment in the possession of items outside one's self. The goal of self-actualization becomes subsumed to be acquisition of possessions. The obstacles to goals of self-actualization created by the material and social conditions of the Special Period make it difficult to recognize the ways in which Cuban society still allows for self-expression—through the arts, for example. In a society wherein the drive for self-actualization is not substituted by market forces, we see the embrace of what Marx posed in the *German Ideology* of the Renaissance person, that is, the fisherman in the morning and the theatre critic in the afternoon. If one cannot find fulfillment by the pursuit of self-improvement—in a system in part influenced by market forces like that of contemporary Cuba—there will be a tendency to seek fulfillment outside one's self and therefore fall into the consumerist trap. The desire for these consumables leads to increased production, increased consumption, and increased use of natural resources for the production of these goods.

In earlier times, officials including Fidel himself have warned against the temptations of consumerism and materialism, but those messages are not being transmitted today. Environmental education could help curb this tendency. Of course, if the state also fails to provide some relief to those seeking fulfillment through employment opportunities and some rewards in terms of housing and basic needs, no public education campaign will be given credence. The government and scientific community, and NGOs such as Nature and Man and Pro-Naturaleza, have been very successful in educating the Cuban about the human impact on the environment. Now they need to make a similar effort with regard to the link between market forces, consumerism, and the environment. Such a campaign cannot be limited to the ideological rejections of consumerism as a by-product of capitalism, for this approach is likely to be met with cynicism. There must be some meaningful dialogue about the challenge of consumerism/materialism as the country moves toward a mixed economy and increasingly integration into a global market dominated by the forces of capitalism and neoliberalism. Ignoring this reality will only erode the legitimacy of the state. Ordinary Cubans can see what is happening in their society, and the longer the top state officials choose to ignore it, the more they will lose. For the humanistic Marx, the moral impetus for the revolution was the fact that capitalism prevented humans from fully realizing their potential. We see this sentiment echoed in Che Guevara's New Man and in revolutionary Cuba's social goals of literacy, free education, universal health care, commitment to the arts, maternity, infant, and elder care, and the eradication of hunger. The revolution placed humans at the center of development; production was increased to fulfill people's needs. These accomplishments need to be protected, as does the country's natural resource base now and into the future no matter what challenges the country faces.

As Antonio Núñez Jiménez put forth, Cubans need to develop a culture of the environment, a sense of environmental ethics and aesthetics so that the laws and policies they are educated about do not simply seem like decrees to follow but connect to their understanding of the human impact on the environment as well as a meaningful connection to nature. Núñez Jiménez correctly anticipated the need to instill in Cubans a love of nature in a similar manner to the love of one's fellow human imperative to the humanist socialist. If that love of nature is inspired in people, they will protect those *mogotes* in Pinar del Río or the *zunzuncito* beyond the laws passed by the Cuban National Assembly. The late Jorge Ramón Cuevas provided Cubans with the interactive respectful manner to foster the understanding of this knowledge and situate the analysis within the experiences of ordinary Cubans.

These lessons as well as the lessons of survival and protection of the revolutionary gains are in the end what will enable the laws of environmental protection to be implemented, expanded, and maintained.

Cuba's environmental framework is indeed ambitious and far reaching. The law includes important embodiments of the principles set forth by the Brundtland Commission and the signators of Agenda 21, the Rio Declarations. Contained within the law is the commitment to social equity, intergenerational equity, intragenerational equity, the precautionary principle, biodiversity, environmental education, control and repair of environmental degradation, and balancing human needs with those of protecting nature.

Three of the most important aspects of the law include 1) the environmental impact review process, 2) the precautionary principles, and 3) the promulgation of the concepts of sustainable agriculture, renewable resources of energy, and environmental education into the framework of environmental law. The environmental impact process necessarily balances economic development for human needs with those of the protecting of nature. It is an exceptionally important aspect in balancing human needs, development, and environmental protection, especially as the country continues to build its mining and tourism industries.

The promulgation of the precautionary principle with heavy emphasis and reliance upon Cuba's wealth of scientific experts and technicians is a crucial step forward for this society and other societies attempting to embark upon sustainable development. When we examine the larger legal and policy precepts, which imbue environmental policy and law around the globe, we note that much environmental regulation is founded on the notion of risk assessment. Indeed what is the risk, what is damage, what is the impact of the damage, and how can we limit the damage via regulation? On the other hand, the precautionary principle focuses on questions of prevention. What is the harm, its impact, and *how* can it be prevented? Some critics may argue that this approach is more difficult to assess given its focus on informed scientific hypothesizing and prediction as opposed to risk assessment based on data collection following environmental damage. But to the proponents of the precautionary principle, it is not always prudent to wait until the damage has been done before ascertaining what the impact might be. In many ways, the focus on precaution by the Cubans and international representatives at Rio places both humans and the natural world more at the center of the analysis. It directs us to discover the causes of pollution or environmental damage and find alternative actions that do not result in such harm. Risk assessment places the emphasis on the activities and parties that might produce the damage. But under a precautionary system, it is the entity, be it public agency or private enterprise, which bears the burden of preventing harm via research, information dissemination about potential damage, monitoring and evaluation, and taking responsibility for any harm done. This changes the dynamic of the way in which these companies or agencies interact both with humans and the environment in that instead of utilizing a *potentially* damaging technology and placing side bets on the occurrence, the company will simply seek the least-damaging technology so as to minimize the risk, thus its own responsibility and liability. The question becomes not how much damage is *acceptable* but how can damage be *avoided*. The precautionary principle as a foundational precept of environmental law is gaining strength around the globe as the European Union recently decided to incorporate it into its environmental laws.

Finally, Cubans have written sustainable agriculture, renewable resources of energy, and environmental education into their environmental law. In a world facing reduced resources of petroleum products for agriculture and energy creation, the transition to more sustainable forms of agriculture and increased use of renewable resources of energy will help Cuba sustain its ability to feed its populace and create the energy required for continued development. Regardless of the strength of a given nation's environmental laws, if the population does not comprehend the human impact of the environment, they will not understand nor will they adhere to those laws. It is absolutely necessary to change the mindset of people so they grasp the importance of why they need to change their behaviors in the home, office, and school so as to lighten their footprint on the planet.

Earlier in the book, we discussed five points that sustainable development must address based on our analysis of the pertinent literature.

Sustainable development must

consider the needs of the people, both basic and transformative;
pay attention to the condition and limits of nature in terms of environmental protection and management;
involve people as active participants in the development process setting forth the needs at the local level;
protect the equity between today's and tomorrow's generation; and
address larger concerns about wealth and land distribution in a society as well as social justice.

Cuba is following a path of sustainable development that follows these precepts and reveals new aspects to be considered by those seeking to build a sustainable society. The island is accomplishing this by a) meeting the basic needs of its citizens in food security, education, and health care; b) facilitating the development of the whole person by its commitments to education, as well as athletic and artistic expression; c) creating a wide-ranging comprehensive environmental protection legal framework; d) developing an environmental consciousness through societal-wide environmental education programs and efforts to develop an environmental ethic; e) developing self-sufficiency in food, energy, and medicine; and f) initiating efforts to enhance public participation in local-based development and planned increases in public involvement in environmental decision making and management.

FINAL LESSONS OF THE CUBAN CASE

The Cuban case demonstrates what people can accomplish in terms of social progress given the political will and the power to do so. Specifically, it is apparent that the adoption of the framework of environmental law and the changes in the relationship between humans and their environment have provided options for balancing economic and social progress while preserving the country's natural resources for future generations. Valuing and meeting basic human needs is a necessary component of a sustainable society. Given increasing awareness of the environmental impact of unmitigated economic growth around the globe and the social and environmental impact of materialism, the revolutionary Cuban example shows what can be done when people, not profits, are the focus.

Cubans often say, "The weapons we have are our ideas." Indeed, other nations, like Venezuela, are already employing Cuban techniques in their own development programs. As we have seen, the Brundtland Report identified five primary areas in which a nation can take steps toward sustainable development: satisfaction of human needs and aspirations, intergenerational equity, a lifestyle within the limits of nature, that is, a model of development that balances human needs with environmental protection, and a legal framework to ensure that these principles are followed. Cubans have made significant advances in all these areas.

The provision of basic needs such as potable water, health care, nutrition, education, and housing helps break the link between human deprivation and environmental degradation. This case study has clearly demonstrated the Cuban commitment to this goal. The development of human capital by supporting people's intellectual and creative growth through educational, athletic, and artistic development is part of the promise of a socialist society, and Cuba receives very high marks in this regard. As for equity between and within generations, because of the long-term commitment to equality that is fundamental to the revolution, and of being on the same island, Cubans conceptualize their situation in relation to others both past and present and these ideas are reflected in the environmental laws the country has adopted. In particular, a lifestyle within the limits of nature will be one based on real needs and aspirations as opposed to those shaped by consumerism and materialism. While Cuba has to some degree addressed this issue with its focus on the development of the intellectual and creative aspects of the person, these efforts could be enhanced by the examination of the dangers of consumerism as part of public education.

A model of development that balances human needs with environmental protection will focus on self-sufficiency, and Cuba's efforts in sustainable and urban agriculture, the development of renewable energy resources, and advances in bioclimatic architecture and green medicine are examples of its progress in this direction. Here the public must be increasingly drawn into the conversation about the nation's development through environmental education and the development of environmental ethics and aesthetics so they will understand why they need to lighten their footprint on the planet and will want to protect the natural world.

Finally, a legal framework for sustainable development is in place, and the challenge is to continue to enforce the laws even in the face of economic difficulty and to expand public participation in the process and transparency to the Cuban public and the international community.

The overarching lesson of the Cuban case is what has been missing from the international discussion of the concept of sustainable development—the social justice aspect. Sustainable development cannot be realized without social justice. The linkages between poverty and environmental degradation are all too clear. We cannot protect social gains and natural resources for future generations unless we protect the ability of the current generation to realize its potential. This means that we must identify the social, political, and economic structures that impede the realization of human potential and counter them with a "culture of nature," social justice, and popular participation.

NOTES

1. Industry experts believe Cuban waters in the Gulf of Mexico might contain substantial deposits, although past explorations have only led to modest discoveries.

Selected References

Acosta, José. "Las leyes de reforma graria en Cuba y el sector privado campesino." *Economía y Desarolllo* 12 (1972): 84–115.

Acosta de la Luz, Lerida. *Proporcionese Salud cultive plantas medicinales.* Havana: Ministerio de Cultura, Edificio Científico-Tecníca (1993).

Acosta Morena, Roberto. "The Environment in U. S.-Cuban Relations: Opportunities for Cooperation." *An Inter-American Dialogue Conference Report,* Washington, D.C: Inter-American Development Bank (1995).

Alarcon de Quesada, Ricardo, President of National Assembly of the Popular Power (ANPP) Introduction to Ley No. 87, Law of the Environment. *Gaceta Oficial,* Havana: Republica de Cuba (July 11, 1997).

Anderson, Robert, Edwin Levy, and Barrie Morrison. *Rice Science and Development Politics: Research Strategies and IRRI'sTechnologies Confront Asian Diversity (1950–1980).* Oxford: Clarendon Press (1991).

Aquilera, Silvio, Especialista dirreción nacional de granos. Ministry of Agriculture. Personal Interview conducted November 29 in Havana, Cuba (1994).

Associated Press. "Cuba makes huge crude oil discovery," (December 26, 2004). Electronic version.

Association of Cuban Geographers. Mesa Redonda Internacional. Río Toa SOS (International Round Table Río Toa conference proceedings. Havana, Cuba, (1998).

Azicri, Max. *Cuba Today and Tomorrow; Reinventing Socialism,* Gainesville: University Press of Florida (2000).

——. "The Rectification Process Revisited," in Halebsky, Sandor and John Kirk (eds). *Transformation and Struggle: Cuba Faces the 1990s.* New York: Praeger (1990).

Barkin, David. *Wealth, Poverty and Sustainable Development.* Mexico City: Editorial Jus: Centro de Ecología y Desarrollo (2000).

Baró Herrera, Silvio. "El Desarrollo Sostenible. Breves consideraciones." Paper presented at *Conferencia Internacional: Etica y Cultura del Desarrollo: Construyendo una economía Sostenible*, Havana, Cuba, May 31–June 5. Baró is senior research at the Center for European Studies, Havana, Cuba (1998).

Beatley, Timothy and Kristy Manning. *The Ecology of Place: Planning for Environment, Economy and Community*. Washington, D.C.: Island Press (1997).

Bebbington, Anthony. *Farmer Organization in Ecuador: Contributions to Farmer First Research and Development*. London: IIED International Institute for Environment and Development (1991).

Bengelsdorf, Carolee, *The Problem of Democracy in Cuba: Between Vision and Reality.* New York: Oxford University Press (1994).

Benjamin, Medea. *The Cuba Reader: The Making of a Revolutionary Society,* Philip Brenner et al (ed.), New York: Grove Press (1989).

Benton, Ted. "Sustainability and Capital Accumulation" in *Fairness and Futurity: Essays on Sustainability and Social Justice,* Andrew Dobson, Oxford University Press: New York (1999).

Bérriz, Luis. "Las energies renovables en Cuba." *Energía y tu*, No. 22, Havana: Cuban Society for Promotion of Renewable Energy and Environmental Respect (CUBASOLAR) (2000).

Bianchi, Andres. *Cuba: The Economic and Social Revolution,* Westport, CN: Greenwood Press (1964).

Blanco, Juan Antonio. "Para Rescatar la Esparanza." Paper presented at *Conferencia Internacional: Etica y Cultura del Desarrollo: Construyendo una economía Sostenible,* Felix Varela Center. Havana, Cuba, May 31– June 5. Blanco is founding president of Center for American Studies in Havana (1998).

Borges Hernández, Teresa. Policy Directorate, Ministry of Science, Technology and the Environment, personal interview conducted January 15 in Havana, Cuba (2003).

Borges Hernandez, Teresa and C. Diaz. "Cuba: política ambiental a tono con los nuevos tiempos." *Temas*, No. 9, Havana (1997).

Boyce, James. *The Political Economy of Growth and Impoverishment in the Marcos Era.* Honolulu: University of Hawaii Press in association with the OECD Development Centre (1993).

British Petroleum. *British Petroleum Statistical Review of World Energy* (2000).

Brudenius, Claes. *Economic Growth, Basic Needs, and Income Distribution in Revolutionary Cuba*. Lund: University Research Policy Institute (1981).

Bustio Ramos, Ania. "Desarrollo sostenible como paradigma de desarrollo. Cuba en camino a la sostenibilidad"Paper presented at *Conferencia Internacional: Etica y Cultura del Desarrollo: Construyendo una economía Sostenible,* Felix Varela Center. Havana, Cuba (May 31–June 5, 1998).

Caguax Communications. *La Agricultura Urbana en la Capital,"* Ministry of Agriculture, Havana, Cuba (1998).

Callicott, J. Baird. "The Land Aesthetic," in *Companion to a Sand County Almanac: Interpretive and Critical Essays*. Edited by J. Baird Callicott. Madison: Wisc.: The University of Wisconsin Press (1987).

Canadian Press. "Sherritt and Pebercan shares rise after news of Cuban oil find," (December 30, 2004). Electronic version.

Cardet Hernando, Luís. "El Programa Alimentario: Su estratégia económica." *Revista Cuba Económica,* 1, no.1, (April–May–June, 1991): 42–54.

Carranza Valdés, Julio, Luis Gutiérrez and Pedro Monreal González. *Cuba: Restructuring the Economy- A Contribution to the Debate*. London: The Institute of Latin American Studies, University of London, (1996).

Castro, Fidel. *History Will Absolve Me*. New York: Lyle Stuart (1961).

———. *Granma Weekly Review*, (Feb. 10, 1986): 2–21.

———. Castro speech in ANPP, 1991a, pp. 117.

———. Speech to the PCC Provincial Assembly, Feb. 5, Fidel Castro Speech Database (FCSDB), 1991b.

———. Castro in interview with *Granma*, Havana; *Granma*. (Sept. 25, 1990).

———. IV Congreso Del Partido Comunista de Cuba, (October 1991).

Chilcote, Ronald H. *Development Theory& Practice: Latin American Perspectives* (Latin American Perspectives in the Classroom), Lanham: Maryland: Rowman and Littlefield (2003).

———. *Theories of Comparative Political Economy*. Boulder, Colorado: Westview Press (2000).

Choncol, Jacques. "El primer bienio de reforma agraria (1959–1961)." Pp. x–x in *Reformas agrarias en America Latina*, Oscar Delgado (ed.). Mexico City: Fondo de Cultural Economica (1965).

Comití Estatal de Estadísticas. *Anuario Estadístico Cubano*, Havana: CEES (1989).

Companioni, Nelso, and Yanet Ojeda Hernández, et al. "The Growth of Urban Agriculture," in *Sustainable Agriculture and Resistance: Transforming Food Production in Cuba*, Fernando Funes et al.(ed), Oakland: CA.: Food First Books, Institute for Food and Development Policy (2002).

Connor, Steve, "Final Proof: Global Warming is a Man-Made Disaster," *The Independent*, London: UK (February 19, 2005).

Conway, Gordon and Edward Barbier. *After the Green Revolution: Sustainable Agriculture for Development*. London: Earthscan (1990).

CUBASOLAR. *Bylaws*. Havana, Cuba, Cuban Society for Promotion of Renewable Energy and Environmental Respect CUBASOLAR (2000).

Cuevas, Dr. Jorge Ramón. Departamento de Lombricultura, Instituto de Suelos y Fertilizantes. Ministerio de Agrícultura. Personal interview conducted November 24 in Havana: Cuba (1994).

———. Personal interview conducted July 16 in Havana Cuba (2000).

———. *An Inter-American Dialogue Conference Report*, Washington, D.C.: Inter-American Development Bank (April 1995).

Deere, Carmen Diana. "Cuba's National Food Program and its Prospects for Food Security" in *Agriculture and Human Values* 10, no. 3 (Summer 1993): 35–51.

———. "Cuba's Struggle for Self-Sufficiency." *Monthly Review* 43, no. 3, (July–August, 1991): 55–74.

———. Testimony to United States House of Representatives Committee on Agriculture, Subcommittee on Foreign Agriculture and Hunger Hearing on "Agricultural Implications of Cuban Trade." Washington, D.C., (May 19, 1994): 1–15.

Del Risco Rodríguez, Enrique. *Cuban Forests: Their History and Characteristics*, Havana: Editorial José Martí (1999).

Desai, B. M. and John Mellor. *Institutional Finance for Agricultural Development: An Analytical Survey of Critical Issues* (Food Policy Review 1), International Food Policy Research Institute (October 1993).

Díaz, Dra. Beatríz and Marta Muñoz. "Biotecnología agrícola y medio ambiente en el período especial cubano." *Cuadernos Africa América Latina*, no 16 (April, 1994): 77–90.

Díaz-Briquets, Sergio and Jorge Pérez-López. *Conquering Nature: The Environmental Legacy of Socialism in Cuba*, University of Pittsburgh Press, Pittsburgh, PA (2000).

Dlott, Jeff et al. "Management of Insect Pests and Weeds." *Agriculture and Human Values* 10, no 3, (Summer 1993): 9–15.

Dunkerley, James. *Power in the Isthmus: A Political History of Modern Central America*. London: Verso (1988).

Edelstein, Joel. "The Evolution of Cuban Development Strategy (1959–1980)." Pp. 225–66 in *From Dependency to Development: Strategies to Overcome Underdevelopment and Inequality*, edited by Heraldo Muñoz. Boulder, CO.: Westview Press (1981).

Environmental News Service. "Cuba Saves Five Internationally Important Wetlands," Environmental News Service, January 6, Online version (2002).

Everson, Frederic. "A Deeper Shade of Green: The Evolution of Cuban Environmental Law and Policy." *Golden Gate University U.L. Review*, 28, Rev, 489 (Spring 1998).

Fabelo, José Ramón. "Socialismo, Mercado y Valores Humano," Paper presented at *Conferencia Internacional: Etica y Cultura del Desarrollo: Construyendo una economía Sostenible*, Felix Varela Center, Havana, Cuba (May 31–June 5, 1998).

Fernández, Olga and José Fabelo. "¿Qué tipo de antropocentrismo: ha de ser erradicado? Paper presented at *Conferencia Internacional: Etica y Cultura del Desarrollo: Construyendo una economía Sostenible*, Felix Varela Center, Havana, Cuba (May 31–June 5, 1998).

Flavin, Christopher. "Rich Planet, Poor Planet." Pp. x–x in *State of the World 2001*, The WorldWatch Institute, New York: W. W. Norton (2001).

Fletcher, Lehman, Eric Graber, William C. Merrill, Erik Thorbecke. *Guatemala's Economic Development: The Role of Agriculture*. Ames: Iowa State University Press (1970).

Freire, Paolo. *Education for Critical Consciousness*, London: Sheed and Ward (1974).

Forster, Nancy "The Revolutionary Transformation of the Cuban Countryside." University Field Staff International Reports. Hanover, NH.: University Field Staff International 26 (1992): 1–18.

——. "Cuban Agricultural Productivity: A Comparison of State and Private Farm Sectors." *Cuban Studies/Estudios Cubanos*11/12, (1981–1982): 10–25.

Foster, John Bellamy. *Ecology Against Capitalism*. New York: Monthly Review Press (2002).

Fuentes, Victor and Manuel Granda. *Conozca las plantas medicinales*. Havana: Instituto Cubano del Libro. Editorial Cienífico-Técníca (1997).

Funes, Fernando. "The Organic Farming Movement in Cuba." Pp. 1–26 in *Sustainable Agriculture and Resistance: Transforming Food Production in Cuba*, Fernando Funes et al. (eds.), Oakland: CA.: Food First Books, Institute for Food and Development Policy (2002).

García Mercedes. "Green Medicine: An Option of Richness." Pp. 212–19 in *Sustainable Agriculture and Resistance: Transforming Food Production in Cuba*, Fernando Funes et al. (eds.), Oakland: CA.: Food First Books, Institute for Food and Development Policy (2002).

García Montaña, Florentin. *Los Aves de Cuba: Especíes Endémícas.* Havana: Editorial Gente Nueva (2002).

García Trujillo, Dr. Roberto. Subdirector. Instituto de Ciencia Animal. Ministerio de Agricultura. Personal interview conducted November 22, 2002 in Havana Province, Cuba.

Gersper, Paul. "Soil Conservation in Cuba: A Key to the New Model of Agriculture." *Agriculture and Human Values,* no. 3 (Summer 1993): 16–23.

Gomez País, Gloria. "Neoliberalismo versus Desarrollo Sostenible," Paper presented at *Conferencia Internacional: Etica y Cultura del Desarrollo: Construyendo una economía Sostenible,* Felix Varela Center, Havana, Cuba (May 31–June 5, 1998).

González Couret, Dania. "Apuntes sobre arquitectura bioclimática," *Energía y tu,* Cuban Society for Promotion of Renewable Energy and Environmental Respect (CUBASOLAR) (2000), No. 22.

González, Rosalia, Yanet Ojeda, and J.L. Pozo. "Impacto social de la Agricultura Urbana," in *Agricultura Orgánica,* 5:2, (agosto) (1999).

González Novo, Teresita and Ignacio García Díaz. *Cuba: Su Medio Ambiente Después de Medio Milenio,* Havana: Editorial Científico-Técnica (1998).

Graña González, Angel. Coordinador General, Fundación de la Naturaleza y el Hombre Antonio Núñez Jiménez (FANJNH). Personal interview conducted July 17 in Mirámar, Havana Province: Cuba (2000).

Granma International. "Cuba researchers receive prize for bird protection," *Granma International,* online edition (March 28, 2002).

——. Wildlife refuge BP TITLE? *Granma International,* online edition (May 28, 2002).

——. Cauto river, 30,000. *Granma International* (September 5, 2002) online edition.

——. Sol Meliá consolidates its position in Cuba, *Granma International* (March 18, 2004). Online edition.

Griffin, Keith. *The Political Economy of Agrarian Change.* Cambridge: Cambridge University Press (1974).

——. Personal communication, Riverside, CA (January 16, 1992).

Gutiérrez Domech, Roberto and Manuel Rivero Glean. *Cuban Nature* Havana: Editorial José Martí (2002).

Halebsky, Sandor and John Kirk (eds.). *Transformation and Struggle: Cuba Faces the 1990s.* New York: Praeger (1990).

Henriquez, Bruno. Personal communication. Cuban Society for Promotion of Renewable Energy and Environmental Respect (CUBASOLAR), Havana, Cuba (June 2003).

Hernández, Barrueta, Dr. Germán. Laborotorio de Rhizobiología Estacion Experimental "La Renee," Ministerio de Agricultura.

——. Personal interviews conducted November 29 and December 1 in Havana, Cuba (1994).

Herrera, Maria, Gricel Alfonso, and Ricardo Herrera. *La Reservas de la Bisofera de Cuba,* Havana: Cuba, Instituto de Ecología y Sistemática, Editorial Academia, Centro Nacional para la Divulgación de la Ciencia y la Técnica (1993).

Herrera, Dr. Ricardo. Department of Ecology and Systematics. Ministry of Science, Technology and Environment. Personal Interview conducted November 23 in Havana, Cuba (1994).

Houck, Oliver Austin. "Cuba's New Law of the Environment: An Introduction." Pp. 1–9 in *Cuban Environmental Law: The Framework Environmental Law and an Index of Cuban Environmental Legislation*, Jerry Speir (ed.), New Orleans: Tulane University (1999).

InterPress News Service. "Cuba Reclaiming Heavily-Polluted Havana Bay," InterPress News Service (September 21, 2000). Electronic version.

Inturralde-Vinent, Manuel "Prologue." *Cuban Nature*, Roberto Gutiérrez Domech and Manuel Rivero Glean (eds.), Havana: Editorial José Martí (2002).

Kay, Cristóbal. "Economic reforms and collectivisation in Cuban agriculture" *Third World Quarterly* 10, no. 3, (July 1988): 1239–66.

Klare, Michael. *Resource Wars: The New Landscape of Global Conflict*. New York: Henry Holt (2001).

Kovaleski, Serge. "With Drugs Scarce, Cuba Tries Natural Cures: Alternative Medicine Clinics Provide Herbs and Homeopathic Remedies," *Washington Post* (March 29, 1999): A10.

Lafferty William and Oluf Laughelle. *Towards Sustainable Development: On the Goals of Development and the Conditions of Sustainability*, St. Martin's Press: New York (1999).

Law Number 33, Protection of the Environment and Rational Use of Resources. *Gaceta Oficial* (February 12, 1981).

Law No. 77, Foreign Investment Law. *Gaceta Oficial* (1995).

Law No. 81, Law of the Environment. *Gaceta Oficial* (July 11, 1997).

Leopold, Aldo. *A Sand County Almanac*. New York: Oxford University Press (1949).

——. *Round River*. New York: Oxford University Press (1953).

Levins, Richard. "The Struggle for Ecological Agriculture in Cuba," *Capitalism, Nature and Socialism* (1990): 121–41.

Lippman, Roger, Tom Lent, Laurie Stone and Cameron Duncan. "Renewable Energy Development in Cuba", *Technology and Society Magazine*, April 1997.

Lockwood, Lee. *Castro's Cuba, Cuba's Fidel*, New York: Vintage Books, 1969.

López, F. "El país espera por la repuesta de los orientales en el año 2000," *Granma*, (January 26, 2000).

Lowe, Jack. "Mirex, Fire Ants and Estuaries," Proceeding of the Workshop on Agrochemicals and Estuarine Productivity (National Oceanic and Atmospheric Administration, Washington, DC, (1982): 65–66.

Lutjens, Sheryl. *The State, Bureaucracy and the Cuban Schools: Power and Participation*, Latin American Perspectives Series No. 17, Boulder, Colorado: Westview Press (1996).

MacEwanArthur. *Revolution and Economic Development in Cuba*. London: Macmillan Press (1981).

Maniates, Michael. "In Search of Consumptive Resistance." Pp.199–236 in *Confronting Consumption*, edited by Thomas Princen, Michael Maniates and Ken Conca, Cambridge, MA: MIT Press (2002).

Marten, Gerald and Patma Vityakon. "Soil Management in Traditional Agriculture" in *Traditional Agriculture in Southeast Asia: A Human Ecology Perspective*, Gerald G. Marten (ed.). Boulder: Westview Press (1986).

Maura, J. A. *Producción de Biopesticidas. El caso de Cuba*. Informe del Taller Regional sobre Tecnologías integradas de producción y protección de hortalizas. Cuernivaca, Mexico: FAO, 1994: 69–74, cited by Pérez y Vásquez, in Funes et al. (2002).

Ministry of Agriculture (MINAG). *Informes annual 1999*. Comisión Nacional de Organopónicos y Huertos Intensivos. Havana: Grupo Nacional de Agricultura Orgánica, MINAG (1999).

Ministry of Agriculture-AgroVide. *Revolución Subterraneo*. Havana, Cuba (1992).

Ministry of Education. *Cuba, organización de la educación 1981–1983; Conferencia Internacional de Educación*. Havana:Cuba: Ministry of Education (1984).

Ministry of Science, Technology and the Environment (CITMA). *Cuba Report to VIII Session of the Commission on Sustainable Development*, New York: United Nations, April. Official English translation (2000).

——. *Situación Ambiental 2002* Ministry of Science, Technology and the Environment (CITMA), (electronic version) (2003).

——. Estrategia Ambiental Nacional, Republica de Cuba (EAN), Ministry of Science,Technology and the Environment (CITMA), Havana, Cuba: CITMA. Reimpression (1999).

——. Agency of the Environment (AMA). *Mission Statement* ,Ministry of Science, Technology and the Environment (CITMA), Havana, Cuba.

——. *Estrategia Nacional Educacíon Ambiental*, p 12, Centro de Información, Divulgación y Educación Ambiental, Agencia de Medio Ambiente, Ministerio de Ciencia, Tecnología y Medio Ambiente (1997).

——. Environmental Mission. Center for Environmental Information, Management and Education (CIGEA), Ministry of Science, Technology and the Environment (CITMA), Havana, Cuba.

——. *Misión Ambiental: Agenda 21 Edición infantile y juvenil de Cuba*, Center for Environmental Information, Management and Education (CIGEA), Save the Children, Ministry of Science, Technology and the Environment (CITMA), Havana, Cuba (2000).

Muñoz, Marta, Centro de Estudios del Desarrollo and Facultad Latinoamericana de Ciencias Sociales (FLACSO). Personal interview conduced November 30 at Universidad de la Habana, Havana, Cuba (1994).

Murphy, Catherine. *Cultivating Havana: Urban Agriculture and Food Security in the Years of Crisis*, Development report 12, Institute for Food and Development Policy, Food First, Oakland, CA (1999).

Murray, Douglas. *Cultivating Crisis: The Human Costs of Pesticides in Latin America*. University of Texas Press (1994).

Naess, Arne. 'Den dypøkologiske bevegelse:aktivisme ut fra et helhetssyn,' In S. Gjerdåker, L. Gule and B. Hagtvet (eds), *Den uoverstigelige grense. Tanke og handling I miljøkampen*. Bergen and Oslo: Chr. Michelsens Institute and J.W. cappelens Forlag (1991) p 37–38, cited in Lafferty and Laughelle (1999): 13.

National Information Agency (AIN), Republic of Cuba. United Nations Qualifies Cuban Model of Urban Agriculture as Very Successful, Havana: Cuba, August 29, 2003.

National Resource Council. *Sustainable Agriculture and the Environment in the Humid Tropics*. Committee on Sustainable Agriculture and the Environment in the Humid Tropics, Board on Agriculture in Science and Technology. Washington, D.C.: National Academy Press (1993).

Navarro, Mireya. "Cuba Passes Law to Attract Greater Foreign Investment," *New York Times*, Section A, pg. 12, (September 7, 1995).

Núñez Jiménez, Antonio. *Hacia una Cultutal de la Naturaleza.* Havana: Cuba, Fundación de la Naturaleza y el Hombre (1998).

——. "Hacia Una Cultura de la Naturaleza"—Cuba: Man and Nature column, *Granma,* (February 14, 1980).

——. *Geografia de Cuba,* Havana: Editorial Nacional de Cuba, Third Edition.

Núñez Jiménez, Lupe Velis de (1965).

——. *Strategic Plan for the Environmental Education Program,* Havana: Cuba, Fundación de la Naturaleza y el Hombre Antonio Núñez Jiménez, (June 1999).

Nuñez Jover, Jorge. "Knowledge, the University, and Socialism." In *Cuba in the 1990s,* José Bell Lara (ed.), Havana: Editorial José Martí (1999).

O'Connor, James R. The Origins of Socialism in Cuba. Ithaca: NY: Cornell University Press (1970).

——. "Is Sustainable Capitalism Possible?" in Martin O'Connor (ed) Is Capitalism Sustainable, New York: Guilford Press (1994).

Oficina Nacional de Estadísticas. Anuario Estadístico de Cuba (2001), Edición (2002), Havana: Oficina Nacional de Estadísticas (2002).

Ojeda, Yanet. "La Granja Urbana: Elemento facilitador del desarrollo del la agricultura urbana." I Fórum Tecnológico Especial de Agricultura Urbana. Nivel Provincial, 1999, cited in Companioni and Ojeda (2002).

Pages, Raisa. "Use of Pesticides Reduced Twenty-fold in Last 15 Years, *Granma International,* (Nov. 23, 2004).

——. "The Worst Drought in over 100 years," *Granma International* (February 25, 2005).

Pastor, Manuel. *External Shocks and Adjustment in Contemporary Cuba,* The International and Public Affairs Center, Occidental College (1992).

Peláez, Orfilo. "Legislación Ambiental: Detener las conductas incorrectas y educar a los violadores," *Granma* (August 15, 2002): 3.

Pérez, Eduardo. Director of Biology and Integrated Pest Management, Instituto de Investigaciones de Sanidad Vegetal. Personal interview conducted in Miramar, Cuba. (December 2, 1994).

Pérez Marin, Dr. Enrique, Ministry of Agriculture and University of Havana, Personal interviews conducted November 18 and November 29 in Havana, Cuba (1994).

Pérez, Nilda, and Luis Vázquez. "Ecological Pest Management." Pp. 109–43 in *Sustainable Agriculture and Resistance: Transforming Food Production in Cuba,* Fernando Funes et al. (eds.), Oakland: CA.: Food First Books, Institute for Food and Development Policy (2002).

Pérez Perera, Dr. Leon Felipe. Cuban Ministry of Science, Technology and the Environment, Personal communication. Ciénaga de Zapata (June 2003).

Portney, Kent. *Taking Sustainable Cities Seriously: Economic Development, the Environment, and Quality of Life in American Cities.* Cambridge, MA: MIT Press (2003).

Princen, Thomas, Michael Maniates and Ken Conca. "Confronting Consumption." Pp. 1–20 in *Confronting Consumption,* edited by Thomas Princen, Michael Maniates and Ken Conca, Cambridge, MA: MIT Press (2002).

Radio Havana. "Cuba Nominated for Special UNESCO award (October 1, 1999). http://www.radiohc.org/espanol.html.

——. "Reduced pollution in Cuba's water reserves," (November 15, 2000) http://www.radiohc.org/espanol.html.

——. "UN recognizes Cuba's Environmental Work," (May 22, 2001).
——. "Havana Bay clean-up is attracting wildlife back to the capital," (June 2, 2001), http://www.radiohc.org/espanol.html.
——. "United Nations Development Program in Cuba receives award," (June 1, 2002). http://www.radiohc.org/espanol.html.
RAMSAR Convention on Wetlands web site. Information Sheet, Cauto Delta Wetland, Ramsar document (September 17, 2002). http://www.ramsar.org.
——. Information Sheet, Buenavista Wetland, Ramsar document, (September 17, 2002). http://www.ramsar.org.
——. Information Sheet, Ciénaga de Lanier Wetland, Ramsar document (September 17, 2002). http://www.ramsar.org.
——. Information Sheet, Gran Humedal del Norte de Ciego de Ávila, Ramsar document (September 17, 2002). http://www.ramsar.org.
——. Information Sheet, Rio Máximo-Cagüe Wetland, Ramsar document (September 17, 2002). http://www.ramsar.org.
——. Information Sheet, Ciénaga de Zapata, Ramsar document (January, 2001). http://www.ramsar.org
Repetto, Robert. *World Enough and Time: Successful Strategies for Resources Management.* New Haven: Yale University Press (1986).
Rey, Gina. "La Habana: Desarrollo sustentable desde la responsabilidad social. Paper presented at *Conferencia Internacional: Etica y Cultura del Desarrollo: Construyendo una economía Sostenible,* Felix Varela Center, Havana, Cuba (May 31–June 5, 1998).
Rey Santos, Orlando. Director, Policy Directorate, Ministry of Science, Technology, and the Environment (CITMA). Personal interview conducted in Havana, Cuba (June 12, 2003).
——. "Reflections on the Legislative Process of the New Environmental Law." Pp. 11–19 in *Cuban Environmental Law: The Framework Environmental Law and an Index of Cuban Environmental Legislation,* edited by Jerry Speir, New Orleans: Tulane University (1999).
Ricoveri, Giovanni. "Culture of the Left and Green Culture." *Capitalism, Nature, Socialism,* 4, 3 (September 1993).
Riera, Lilliam. "Almost a quarter of the island forest covered," *Granma International* online edition (January 10, 2005).
Rivero Glean, Manuel. *Conoza Cuba: Flora y Fauna.* Havana: Editorial José Martí (1999).
Rodríguez Nodals, Adolfo, Elizabeth Peña Turruella, Nelso Companioni Concepción, and Miriam Carrión Ramírez. "La Agricultura Urbana en Cuba: Su Contribución a la Alimentación," Instituto de Investigaciones Fundamentales en Agricultura Tropical "Alejandro de Humboldt." INIFAT, Ministerio de la Agricultura, presented at the IV Convención Internacional sobre Medio Ambiente y Desarrollo, del 2 al 6 de junio, La Habana, Cuba (2003).
Roque Molina, Martha G. "Una propouesta de periodización del desarrollo histórico de la Educación Ambiental (parts 1–2)", *Cuba: Medio Ambiente y Desarrollo,* CIGEA/CITMA, Ano 2, No.2/2002 (electronic version). 2001 "La educación ambiental: Acerca de sus fundamentos teóricos y metodológicos, *Cuba: Medio Ambiente y Desarrollo,* CIGEA/CITMA, Año 1, No. 1 (electronic version).

Rosset, Peter. Director, Food First, San Francisco. Personal phone interview conducted on January 20, 1995.

Rosset, Peter, and Medea Benjamin (eds). *The Greening of the Revolution: Cuba's Experiment with Organic Agriculture.* Victoria, Australia: Ocean Press (1994).

Rostow, Walt. *The Stages of Economic Growth: A Non-Communist Manifesto.* Cambridge: Cambridge University Press, 1960.

Sanchez, Ricardo. Concluding Speech. Remarks presented at IV Convención Internacional sobre Medio Ambiente y Desarrollo, "Un Mundo Mejor Es Posible," Havana, Cuba, June 6. (author's notes) (2003).

——. "Commentary". Remarks presented at *Conferencia Internacional: Etica y Cultura del Desarrollo: Construyendo una economía Sostenible,* Felix Varela Center Havana, Cuba May 31–June 5. Sanchez is vice-minister of the Cuban Ministry of Science, Technology and the Environment, CITMA (1998).

Saouma Edouard in Hope Shand. *La diversidad de la naturaleza un patrimonio valioso,* Rome, Italy: FAO, 1993, quoted in Cuevas 1995: 30.

Schwartz, Rosalie. *Pleasure Island: Tourism and Temptation in Cuba.* Lincoln, Neb.: University of Nebraska Press (1997).

Shishkoff, Nina. "Plant Diseases and Their Control by Biological Means in Cuba." *Agriculture and Human Values* 10, no.3, (Summer 1993): 16–23.

Sklar, Barry. "Cuban Exodus 1980: The Context," Pp. x–x in *The Cuba Reader: Making of a Revolutionary Society,* Philip Brenner et al. (eds.), New York: Grove Press (1989).

Stone, Lori. "PV Powered," *Home Power,* no. 86, December 2001–January 2002.

Stowell, Michael. "The Remarkable Mother of Invention," *Swans,* Arcata, CA, May 14, 2001.

Sweezy, Paul. "Capitalism and the Environment." *Monthly Review,* 41; (June 2, 1989).

Thrupp, Lori Ann and Tenelfe Pérez. "Breaking chemical Dependency: The Remarkable Case of Cuba." Unpublished paper (1989).

Trim, Alan. "Acute Toxicity of Emulsifiable Concentrations of Three Insecticides Commonly Found in Nonpoint Source Runoff into Estuarine Waters to the Mummichog, Fundulus heteroclitus" in *Bulletin of Environmental Contamination and Toxicology* Vol. 38 (1989): 681.

Trinchet, Oscar. "Rendimientos agrícolas de algunos cultivos." *Cuba Económica* (1990): 140–57.

UNEP. Alejandro de Humboldt National Park information sheet, World Heritage Sites, UNEP, http://www.unep-wcmc.org/sites/wh/alejandro.html.

——. UNEP Press Release, *Global 500 Roll Call,* May. Electronic version (2001).

UNESCO, Man and the Biosphere (MAB), *Península de Guanahacabibes,* www.unesco.org/MAB.

——. *Cuchillas del Toa Biosphere Reserve Information,* UNESCO, www.unesco.org.

——. *Baconao Biosphere Reserve Information,* UNESCO, www.unesco.org/MAB.

U. S. Department of Energy, Energy Information Administration. *International Energy Outlook 1999,* Washington, D.C.: DoE/EIA, Table A4, 1999.

Uriarte, Miren. *Cuba: Social Policy at the Crossroads: Maintaining Priorities, Transforming Practice,* Oxfam America (2002).

Webster, Andrew. "Cuba does what comes naturally in medical care: U. S. trade embargo forces reversion to centuries-old healing techniques," *Globe and Mail*, June 29, 1999.

Weir, David and Mark Shapiro. *Circle of Poisons: Pesticides and People in a Hungry World*. San Francisco: Food and Development Policy Institute (1981).

World Commission on Environment and Development. *Our Common Future*. Oxford: Oxford University Press, 1987.

World Health Organization. *Public Health Impact of Pesticides Used in Agriculture*. Geneva: WHO (1990).

Zimbalist, Andrew. "Teetering on the Brink: Cuba's Post-CMEA Economic and Political Crisis." *Journal of Latin American Studies* (1993).

Index

About the Author

Pamela Stricker, Ph.D. is an assistant professor in the political science department and a founding faculty member in the environmental studies program at California State University, San Marcos. She serves as a coordinating editor for the peer-reviewed journal, *Latin American Perspectives*. Stricker also serves on the Council of the Science, Technology and Environment subsection of the *American Political Science Association*.